AF249021

PHalarope Books are designed specifically for the amateur naturalist. These volumes represent excellence in natural history publishing. Most books in the PHalarope series are based on a nature course or program at the college or adult education level or are sponsored by a museum or nature center. Each PHalarope book reflects the author's teaching ability as well as writing ability. Among the books:

The Amateur Naturalist's Handbook
Vinson Brown

Botany in the Field: An Introduction to Plant Communities for the Amateur Naturalist
Jane Scott

A Field Guide to the Familiar: Learning to Observe the Natural World
Gale Lawrence

Insect Life: A Field Entomology Manual for the Amateur Naturalist
Ross H. Arnett, Jr., and
Richard L. Jacques, Jr.

A Natural History Notebook of North American Animals
National Museum of
Natural History, Canada

Owls: An Introduction for the Amateur Naturalist
Gordon Dee Alcorn

The Plant Observer's Guidebook: A Field Botany Manual for the Amateur Naturalist
Charles E. Roth

Pond and Brook: A Guide to Nature Study in Freshwater Environments
Michael J. Caduto

The Sky Observer's Guidebook
Charles E. Roth

Suburban Wildflowers: An Introduction to the Common Wildflowers of Your Back Yard and Local Park
Richard Headstrom

Suburban Wildlife: An Introduction to the Common Animals of Your Back Yard and Local Park
Richard Headstrom

Thoreau's Method: A Handbook for Nature Study
David Pepi

The Wildlife Observer's Guidebook
Charles E. Roth, Massachusetts
Audubon Society

Wood Notes: A Companion and Guide for Birdwatchers
Richard H. Wood

HEAVEN
AND
EARTH

HEAVEN
AND
EARTH

*A Fieldbook
of
Foolproof Nature
Identifications*

THOMAS CARL PETERSON

Illustrated by Susan Maffet Peterson

PHalarope
Books

PRENTICE HALL PRESS · New York

Copyright © 1986 Prentice-Hall, Inc.
All rights reserved, including the right of reproduction
in whole or in part in any form.

Published by Prentice Hall Press
A Division of Simon & Schuster, Inc.
Gulf+Western Building
One Gulf+Western Plaza
New York, NY   10023

A PHalarope Book

PRENTICE HALL PRESS is a trademark of Simon & Schuster, Inc.

Library of Congress Cataloging-in-Publication Data

Peterson, Thomas Carl.
  Heaven and earth.

  Includes index.
  1. Natural history. 2. Animals—Identification.
3. Plants—Identification. I. Title.
QH45.5.P47 1986   508   86-5902
ISBN 0-668-06301-7

Manufactured in the United States of America

Designed by Stanley S. Drate/Folio Graphics Co. Inc.

10 9 8 7 6 5 4 3 2 1

First Edition

Dedicated to
Marian Malmquist Peterson

Acknowledgments

I would like to thank the many amateur and professional naturalists from coast to coast who reviewed this work. The comments and suggestions from Rick Bantz, Art Compton, Sydney Jacobs, George Johnson, Caryl Morton, Jim Pease, Sally Portman, and Ken White improved the book. I would also like to thank Charles A. Whitney for his assistance with Venus's dates, and Henry Rasof for his helpful advice on format and style.

—THOMAS CARL PETERSON

I would like to thank Stephen Lang, Dennis Kirschbaum, and the University of Wisconsin-Madison Departments of Entomology and Wildlife Ecology for providing slides and specimens without which the illustrations could not have been made.

—SUSAN MAFFET PETERSON

Contents

Identification Keys   xv
Introduction   xvii

REPTILES AND MAMMALS

Poisonous Snakes   2
Thirteen-lined Ground Squirrels   4
Bats   6
Dog and Cat Tracks   8
Rabbits   10
Opossums   12
Raccoons   14
Muskrats, Beavers, and Nutrias   16

PLANTS

Coontail   20
Duckweed   22
Waterlilies and the American Lotus   24
Rushes and Bulrushes   26
Sedges   28
Scouring Rush   30
Horsetail   32
Puffball Mushrooms   34
Spruce and Pine Trees   36
Bloodroot   38

Jack-in-the-pulpit 40
Mayapple 42
Wild Columbine 44
Stinging Nettle 46
Velvetleaf 48
Chicory 50
Queen Anne's Lace 52
Prickly Pear Cactus 54
Common Mullein 56
Common Plantain 58
Goldenrod 60

INSECTS

Goldenrod Galls 64
Spittlebugs 66
Dragonflies and Damselflies 68
Mayflies 70
Water Striders and Whirligig Beetles 72
Centipedes and Millipedes 74
Daddy Longlegs 76
Black Widow Spiders 78
Black-and-yellow Garden Spiders 80
Sulphur Butterflies 82
Swallowtail Butterflies 84
Aphids 86
Tent Caterpillars and Fall Webworms 88
Banded Woolly Bear Caterpillars 90

WEATHER SIGNS

Spring Air Wars 94
Clouds 96

Highs and Lows   98
Prevailing Westerlies   100
Warm Fronts and Cold Fronts   102
Lunar Halos   104

★ NIGHTTIME PHENOMENA

The Big Dipper and the Little Dipper   108
Orion   110
The Pleiades   112
The Milky Way   114
Venus   116

BIRDS

Barred Owls and Great Horned Owls   120
Nighthawks   122
Chimney Swifts   124
Mourning Doves   126
American Kestrels   128
Turkey Vultures   130
Red-tailed Hawks   132
Crows, Hawks, and Ravens   134
Belted Kingfishers   136
American Coots   138
Mallards   140
Canada Geese   142
Killdeers   144
Brown-headed Cowbirds   146
Horned Larks   148
Meadowlarks   150
American Goldfinches   152
Chickadees   154

Juncos   156
Nuthatches   158
Downy and Hairy Woodpeckers   160
Redheaded Woodpeckers   162
Northern Flickers   164
Waxwings   166
Eastern Kingbirds   168
Wrens   170

HELPFUL TOOLS

Binoculars and Other Tools   174

Index   179

Identification Keys

The symbols below, found in the right-hand margin of each text page, indicate subject categories and the pages on which they are to be found.

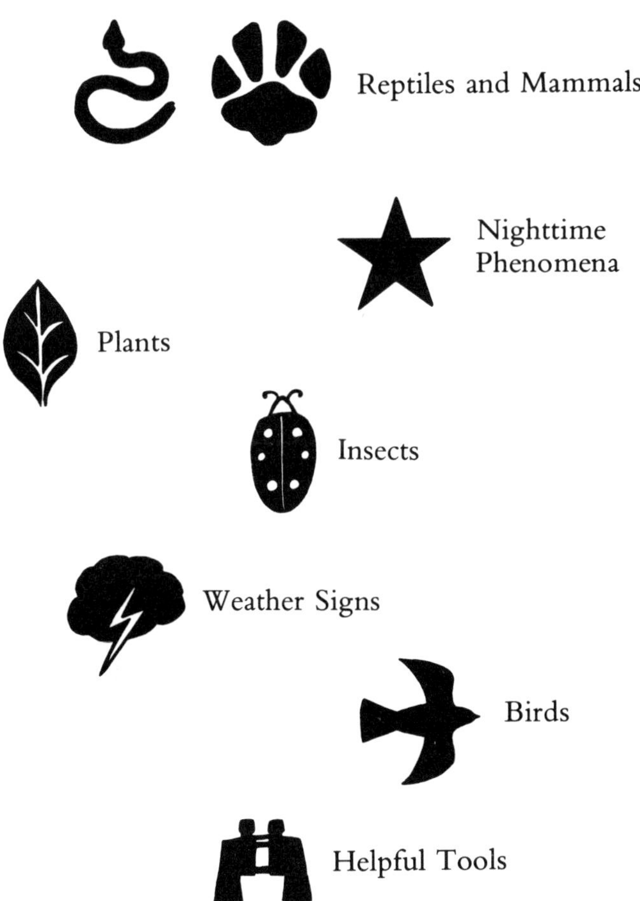

Introduction

How many constellations do you know? If you don't know any, this book will quickly teach you a few. If you do know some constellations, you'll remember that a few of them, like the Big Dipper, are easy to learn while most constellations are hard to learn. Well, the same thing is true about almost all other aspects of nature: You can very easily learn to identify certain wildflowers, for instance, even though most wildflowers are difficult to learn.

Heaven and Earth covers almost all aspects of nature: reptiles, mammals, plants, insects, weather signs, constellations, and birds. But to be included in the book, the subjects had to meet two criteria. First, each had to have some special characteristic that provides a foolproof means of identifying it. Second, each had to be common enough to be seen frequently.

Learning to name a creature or phenomenon is nice, but understanding why the creature behaves the way it does or how the natural phenomenon is created is much better than just being able to name it. It is this understanding that breathes life into the identifications.

Wherever possible, the book describes in simple terms the physiological, ecological, psychological, behavioral, mechanical, or structural aspects that make that particular identification easy. For example, knowing how the rabbit swings its hind legs before and outside its front legs as it runs is the key to identifying rabbit tracks. Historical references and old wives' tales also sometimes contain relevant bits of information or lore that can assist us in remembering an identification. For example, knowing that runaway slaves used to "follow the Drinking Gourd" north to freedom makes finding the Little Dipper, with the North Star in its handle, much easier, because this constellation really doesn't look much like a dipper.

Many rules of thumb for identifying natural phenomena are mentioned in the book. These rules of thumb are simple, easy to use, and widely applicable. Unfortunately, they are not 100 percent accurate. Nevertheless, naturalists rely on them all the time, and you can too. Just keep in the back of your mind that you might occasionally run into an exception to the rule.

The reasons for the exceptions vary with each rule and type of subject. Many weather systems, for instance, vary enough to create their own exceptions. Turkey vultures can be distinguished from other birds by the way they soar with their wings held in a shallow V—except that the zone-tailed hawk (a fairly rare tropical hawk that ranges north into parts of Texas, New Mexico, and Arizona during the summer) also soars with its wings in this configuration.

You can start reading the book at the beginning or start with any subject that catches your interest. When you've read it all, you'll have a well-rounded knowledge of many fields of nature study. If you are working on a science or nature project involving insects, for instance, you'll find the insects, all of which are easy to identify, together in one part of the book. If you are going camping in the woods and you want to learn about woodland plants, you'll find the plants grouped together and arranged by habitat, so that woodland plants appear in one section. If your scout troop is going canoeing or exploring around a lake, you can take the book with you and look up the easy-to-identify aquatic plants and then flip over to the aquatic birds, aquatic mammals, or aquatic insects.

Most of the creatures you'll read about can be found throughout the continental United States and southern Canada. Some of the subjects have very broad ranges and can be found in Alaska and Hawaii, too. Others have considerably smaller ranges, and these are mentioned in the text.

Whether you live in southern California or northern Maine, in a city, a forest, or a desert, the nature subjects described in this book are all around you. Even in a city, clouds fly past, birds sing in the trees, and wild plants grow in untended corners. Studying a small wildflower growing in a city can be just as interesting and give you the same sense of peace and wonder as studying a wildflower growing in the wilderness.

I hope that the next time you see chicory flowering in a city or Orion shining in the night sky, this book will let you greet them like old friends.

REPTILES
AND
MAMMALS

Poisonous Snakes

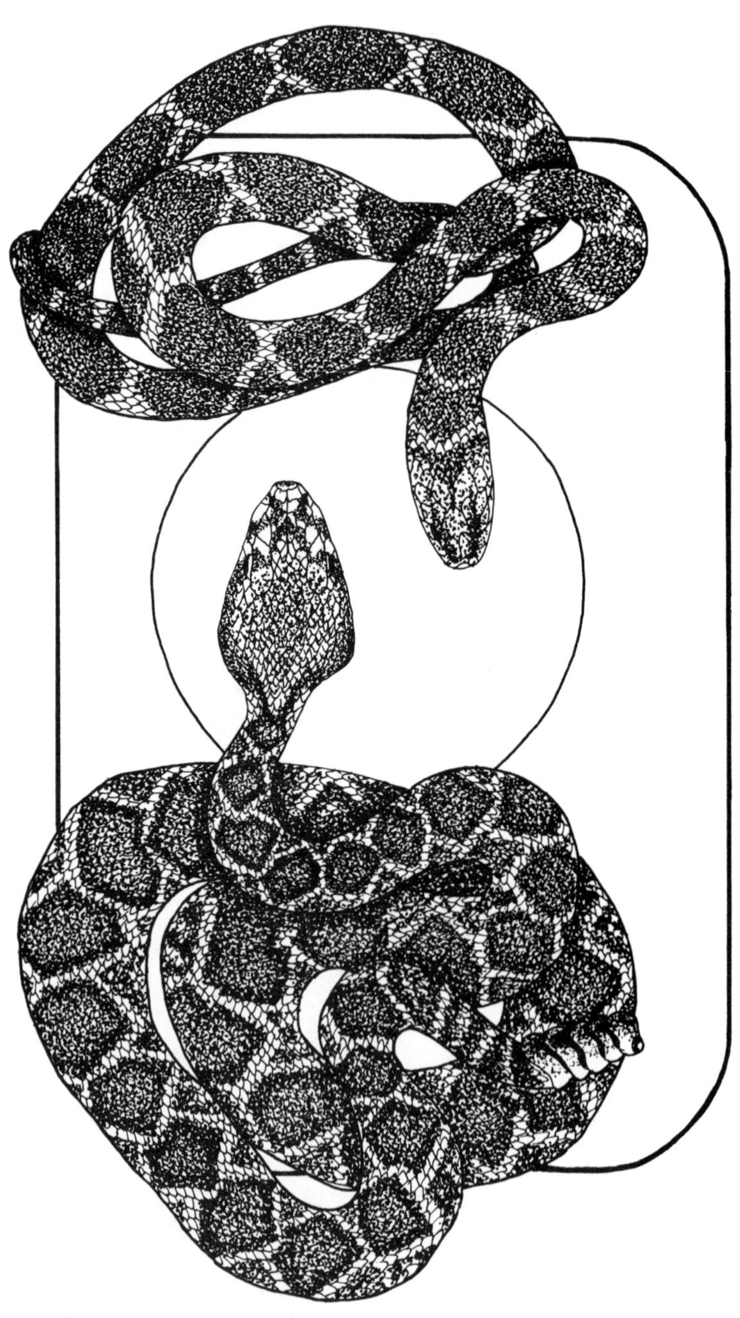

Poisonous snakes can be found in most of the United States and at least portions of southern Canada. If you come upon a snake you don't recognize, here is a quick test to tell whether it is poisonous or not: Is the head larger than the neck? Most nonpoisonous snakes have such a gradual change from the neck to the head that you'll find it hard to tell just where the head starts. But, except for the coral snake, all poisonous snakes in this country have heads distinctly wider than their necks. A few nonpoisonous snakes, such as the hognose snake, also have heads larger than their necks, but it is better to err on the side of caution.

Coral snakes, the poisonous exception to this head-neck identification rule, are colored with rings of red, yellow, and black, with the black bands always between the yellow. There are a couple of similarly colored nonpoisonous snakes, such as the scarlet king snake, but their black bands aren't surrounded by yellow. Again, it is better to err on the side of caution and give all red-yellow-and-black-banded snakes a wide berth; besides, nonpoisonous does not mean harmless.

Two types of coral snake live in the United States, one in the Southwest, mainly in southeast Arizona and western New Mexico, and the other mainly in the Southeast from North Carolina through southern Texas, though it can also be found in a few other locations, as far north as Indiana.

The other poisonous snakes of North America, those whose heads *are* wider than their necks, are pit vipers. Pit vipers are characterized by plump bodies, short tails, and broad triangular heads. Copperheads, cottonmouths (water moccasins), and rattlesnakes all have small heat-sensing pits between their nostrils and eyes. These enable the snakes to accurately strike any prey, such as a warm-blooded mouse in the grass, by distinguishing its body temperature from its colder surroundings. Unfortunately, these pits are too hard to see to be of any value in identifying live snakes.

Rattlesnakes grow an extra rattle button (a scaly, buttonlike appendage) at the end of their tails every time they molt, and they may molt four times a year. The presence of a rattle will positively identify a rattlesnake, but you should not rely solely on rattling for rattlesnake identification. Rattlesnakes don't always give warning rattles, and some nonpoisonous snakes will shake their tails, rattling leaves. Furthermore, a snake may have lost its rattles in accidents with cars, farm machinery, or even rocks.

Thirteen-lined Ground Squirrels

Thirteen-lined ground squirrels are frequently confused with chipmunks, but there are two easy ways to tell them apart. Thirteen-lined ground squirrels run with their tails down, while most chipmunks run with their tails up, and chipmunks tend to live in brushy or woodsy areas, while thirteen-lined ground squirrels live in areas with short grass, such as prairies and golf courses.

You'll find thirteen-lined ground squirrels in the central part of North America. They once lived mainly in the dry shortgrass prairies just east of the Rocky Mountains. Thanks in part to mowed highway rights-of-way and other man-made shortgrass environments, thirteen-lined ground squirrels have extended their range. They can now be found as far west as Alberta, Canada, and New Mexico and as far east as Ohio. They are also quite common in the Midwest, but haven't moved into the Southeast at all.

Thirteen-lined ground squirrels are about the same size as chipmunks, averaging five or six inches long, not including their three- or four-inch tails. Their backs and sides are covered with alternating brown and whitish lines and rows of whitish dots running lengthwise. Their bellies are yellowish white. Their small ears and short legs are well suited for running down underground tunnels.

You can look for these squirrels near their burrows in all sorts of mowed grassy areas, including city parks. The best time to spot them is on warm days. You'll often see them standing on their hind legs, erect and alert. They may stand perfectly still watching you approach, then suddenly dash to their holes, keeping a low profile as they run. Once safe inside their burrows, they'll often poke their heads out again and keep watch.

To find a thirteen-lined ground squirrel burrow you have to look carefully when a ground squirrel disappears into one, because these holes are not marked in any way. They are usually only one or two feet deep at most points, but may be as long as twenty feet and contain a food storage room, separate sleeping quarters, a hidden exit hole, and a deeper section, where the ground is warmer during severe winter, for hibernating.

Thirteen-lined ground squirrels eat just about anything a prairie or other shortgrass habitat has to offer except grass leaves. About half their diet is plant matter, such as seeds, roots, and fruit, and the other half is animal matter, such as grasshoppers, caterpillars, and beetle larvae.

Bats

It's the Fourth of July and you're waiting for the sky to darken enough for the fireworks to start. As you gaze up at the pale sky, you see a small bird flying by. But you notice that its flight seems jerky compared with that of most birds, and its wings are almost as wide as the body is long. In fact, it's not a bird at all—it's a bat.

You can see bats just about anywhere there are the flying insects that they eat, from city streets to desert wildernesses. Bats are easiest to see silhouetted against the sky, so look for them on any nice summer evening just before the last light has left. Probably the best place to look for bats is the shore of a pond. Ponds and other bodies of water usually attract bats because they have a lot of flying insects around them and are convenient for drinking.

During the day bats rest in many different places. They'll hang upside down in among the leaves on trees, in caves, in hollow trees, and in little and large nooks on buildings. Church belfries are such commonly used roosting nooks that a person who is a little crazy, or "batty"—eccentric, like a bat's flight pattern—is sometimes said to have bats in his belfry. A few species of bats migrate south for the winter, but most of North America's bats hibernate, in caves.

Bats "see" at night using echolocation: a system of navigation in which they emit high-frequency sounds and listen to the echoes that bounce back from the objects around them. These sounds are too high-pitched for humans to hear, which is just as well, as scientists say bats call out very loudly. Some moths can hear bats, though, and when they do they drop to the ground or fly very erratically in an apparent attempt to evade the bat. Bats can "see" with echolocation during the day, too, but flying during the day entails the risk of getting caught by hawks, so bats normally do not start flying until well after the sun has set.

One of the few small birds still flying at twilight is the swift. Bats are about the same size as swifts; the typical North American bat is about four inches long, with a wingspan of about ten inches. But you can easily tell swifts and bats apart (see page 124).

Dog and Cat Tracks

Dog and cat tracks are probably the most common tracks you'll see. Following the track of a bobcat or a domestic cat, a coyote or a neighbor's dog, can teach you a lot about your local environment. In urban and suburban areas, domestic cats are one of the chief predators of wildlife, and their tracks can take you to where mice and other small creatures live. Following dog tracks can lead you to animal burrows and other signs of wildlife you wouldn't have seen otherwise and certainly don't have the nose to sniff out yourself.

You can identify cat tracks in snow or mud by the imprint of the pad on the bottom of their paws, four toe prints, and no claw marks. Cats walk on only four toes, and they keep their claws retracted when they walk so the claws will stay sharp for catching prey. Many animals with four legs can see where their front feet will go and automatically put their hind feet in about the same place. Cats do this a lot, which sometimes makes their tracks blurry and can force you to look closely at the track to identify it. Bobcat prints are about two inches long. Mountain lion and lynx prints are three to four inches long.

Dogs also have tracks with a pad and four toes. You can identify the tracks of dogs and their cousins the foxes and coyotes by the pad, four toes, and claw marks. Dog claws are not retractable, so they will almost always make claw marks, which show up as little dots in front of the toe prints. You may have to get down on your hands and knees and look very closely at a number of prints to be sure if there are claw marks or not. Wolf prints are 4 to 5 inches long, coyote and red fox prints are about 2½ inches long, and gray fox prints are about 1½ inches long.

Wildlife researchers studying lynx, for example, will often drive down small roads in the morning after a fresh snow, checking out all the tracks that cross a road until they find a lynx track. They will then follow the track (backward to make sure their presence won't alter the lynx's behavior), recording such information as how often the lynx sat or laid down. Where the lynx leapt after prey, they record the number of leaps, and how long the leaps were. Judging from blood, feathers, tufts of fur, or other tracks in the snow, they determine what prey the lynx was after, and whether it was successful.

You can do much the same research when following local fox or wild or domestic cat or dog tracks.

Rabbits

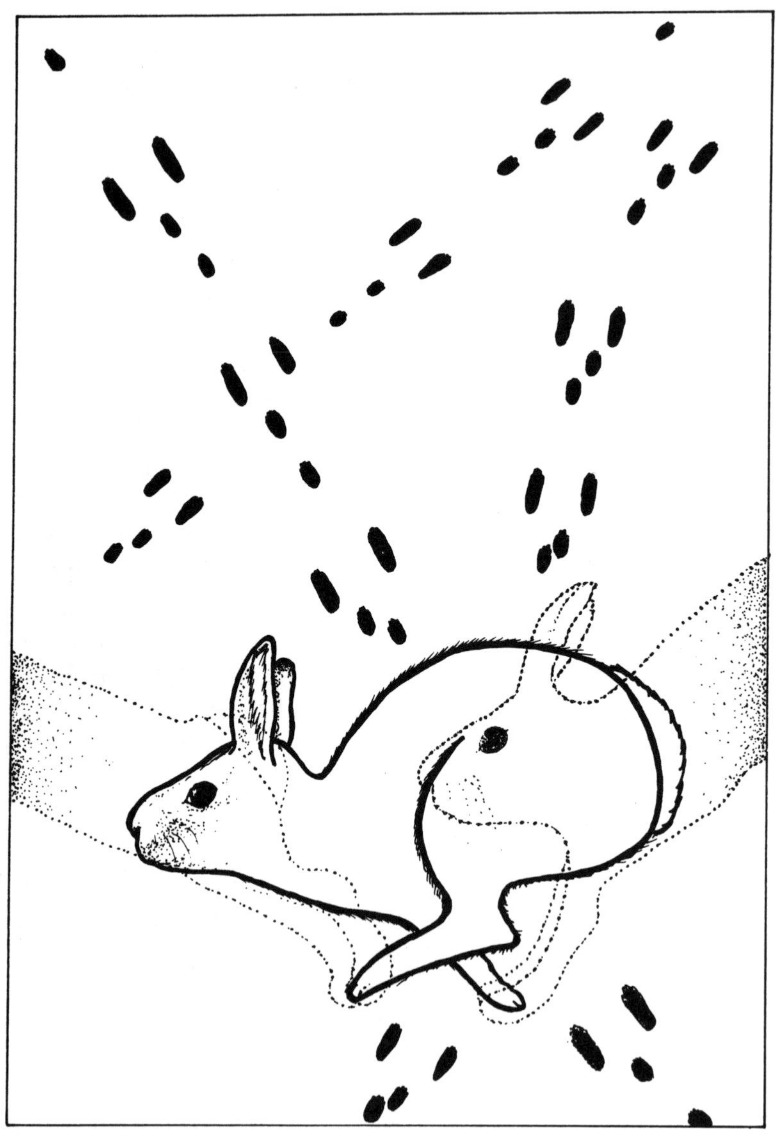

Rabbit tracks can tell you at a glance which way the rabbit was going and how fast. Identifying rabbit tracks is almost as easy as identifying rabbits once you know how rabbits run.

We all know that most rabbits walk and run by hopping. Coming down from a hop, rabbits land on their front feet first. The hind feet are then swung forward so they actually land in front and outside of the front feet. Therefore, rabbit tracks show two small prints close together and two larger hind-feet prints farther apart, followed by a space and another set of prints. The hind feet always land pretty evenly, while the front feet usually land one ahead of the other.

The wide prints are always in front of the narrow prints, and point to the direction in which the rabbit was traveling. You can judge how fast the rabbit was going by the length of the space between sets of prints. If the tracks are about one foot apart, the rabbit was walking. Three feet would be a gentle run. Anything over eight feet should be considered a mad dash.

The best time to look for rabbit tracks is first thing in the morning after a light snow. You can learn a lot about rabbits, and about nature in general, by following rabbit tracks. You may notice that the rabbit you're following ran across an open field and slowed down when it got near cover. This is because rabbits are more likely to be caught by predators such as great horned owls when they are out in the open.

Keep following and the rabbit will often lead you to a narrow little path that winds through shrubbery and underbrush. Even after a fresh snow, there may be many sets of prints running down the rabbit trail. Following the trail may take you to the rabbits' den. You can assume that any hole in the ground or brush pile that the trail leads into has some rabbits in it.

The rabbit track might also take you to where the rabbit has been eating. You might notice trees or shrubs that rabbits have eaten the bark from, and be able to tell which types they like best. They also eat plants such as clover. You'll also notice rabbit droppings deposited where you'd expect rabbits to feel pretty safe, such as in a brier patch. Rabbit droppings are easy to identify. They are round little balls about half an inch in diameter.

Opossums

You can easily identify the opossum by its white face, pointed nose, and long ratlike tail, which measures between ten and twenty-one inches. Its body will be light gray to dark brown or black, and about the size of the average house cat. But it is a shy, nocturnal creature, so you may not even know one lives nearby until you see its tracks. The key to identifying opossum tracks is the inside, or "big," toes on their hind feet, which stick out much like our thumbs. The tracks made by their hind feet are about two inches across or a little larger and show all five toes, with the "big" toe swung so far off to the side that it points inward or even backward. Front-feet tracks also show five toes. The "big" toes on the hind feet never leave claw marks, but all the other toes may. Also, you will frequently see the mark left by the ratlike tail dragging on the ground.

Opossums live mostly in woodlands and farmlands with some trees. Look for their tracks crossing dirt roads, in the mud along stream banks, or in fresh snow. Opossums live mainly on the West Coast from southern British Columbia to southern California and in the eastern half of the United States from Texas and Colorado to Florida and Massachusetts and north to mid-Minnesota and southern Ontario. Northern opossums stay in their dens ("den up") during the worst part of the winter, but they do not hibernate.

Climbing trees is the opossum's main defense against predators. They wrap their hairless tail around branches to help them climb. (In winter, sometimes part of it is frozen off on foraging trips.) If they can't make it to a tree in time—and they often can't, because they run so slowly—their next defense is to hiss, bare their teeth, and prepare to fight. When all else fails, they may suddenly "play possum" and appear to die. "Playing possum" is not good acting, but rather some sort of fit or seizure. The fact that playing dead sometimes keeps them from being eaten implies that opossums must smell bad to coyotes or other predators when in this state.

Opossums raise their young in a pouch, much as their distant relatives the kangaroos do. As the young get too big for the pouch, they may ride on the mother's back as she forages for food, such as fruit, corn, mice, and dead animals.

Raccoons

Out walking through the woods, you are about to cross a stream when you notice some clear tracks in the soft mud of the stream bank. At first glance they look much like a child's hand print showing all five fingers and a child's footprint showing all five toes. The footprint is only about four inches long, about the size of a one-year-old child's foot. Looking closer, however, you notice that the toes are too long and skinny for a child's foot. And all the toes and fingers show claw marks in front of them. They are raccoon tracks.

The best places to look for raccoon tracks are the banks of wooded streams, lakes, and other bodies of water. Raccoons come down to the water at night to hunt for crayfish, frogs, and other small creatures, which they catch with their dexterous paws. Raccoons eat many other things as well, including fruit, nuts, corn, baby birds, small land animals such as mice or squirrels, and human garbage. So you can also find raccoon tracks in many other places, such as fields, roadsides, and around houses that aren't too far from the raccoon's den.

Raccoons prefer to make their dens in hollow trees in woodlands, where they can quickly climb trees to get away from their two main predators: dogs and men. They will, however, also live in caves and burrows originally dug by other animals, so they are not limited to woods.

Southern raccoons are active all year long. In the North, raccoons sleep during the coldest part of the winter, though they don't actually go into hibernation (a special deep sleep during which breathing, heart rate, and temperature all decrease drastically). When the weather moderates, they will often resume their nighttime foraging.

Sometimes you'll see raccoons at night in the light of flashlights or car headlights and their normally black eyes will glow green or reddish. Like many other nocturnal animals, raccoons have a reflective layer in their eyes behind the layer of photoreceptive cells that see light. This improves their vision in dim light because they are able to pick up reflected light.

In areas such as the edge of towns or park campgrounds where raccoons are not hunted and can find food around people, they become bolder and, come evening, may walk out in plain sight. You can easily identify these grayish-brown, medium-size animals by the black masks around their eyes and the rings on their bushy tails.

Muskrats, Beavers, and Nutrias

You're sitting on the shore of a lake watching the sun set and you notice a furry animal swimming quietly by on the surface of the water. What animal is it? North America only has three mammals that are commonly seen swimming quietly on the water surface: muskrats, beavers, and nutrias. You may see them during the day, but all three are most likely to be seen in the half-light of sunrise or sunset. Muskrats and beavers can be seen on streams, rivers, lakes, and ponds throughout most of North America. Nutrias are South American imports that escaped from game farms in the United States during the 1940s, and their range is still pretty much restricted to the Southeast, Texas, Maryland, northeastern Oregon, and eastern Washington.

You can easily distinguish muskrats from beavers by size. Muskrats are little guys about ten to twelve inches long, not including their long ratlike tails, and weigh one to four pounds. Beavers are big, about two or three feet long, not including their tails, and may weigh thirty to sixty pounds. Nutrias are in between beavers and muskrats in size, so if you are in an area where nutrias live, identifying these three animals is trickier. You'll need to look closely at the animal's size and also look for other signs indicating which animals live there.

The most common sign of muskrats is their lodges. They look like domes or piles of mud and marsh plants one to three feet high built in shallow water near a marsh. They have an underwater entrance that allows muskrats to swim out under the ice in winter and eat the leaves and roots of plants, such as waterlilies and cattails, growing in the water.

If beavers are around you'll see the unmistakable sign of trees being chewed through, especially willow and aspen trees growing along the shore. Beavers are famous for building dams and the stick-and-mud lodges in which they live, but neither may be present. The dams may not be needed, and the beavers may dig a burrow in the riverbank instead of building a lodge.

Nutrias are so well adapted to aquatic life that they can even nurse their young while swimming. They like the safety of water around them. But they apparently like to be out of water when eating cattails and other plants. So the most common sign that nutrias are in the area are their feeding platforms—small islands of aquatic vegetation about five or six feet wide.

PLANTS

Coontail

Coontail is one of our most common aquatic plants. You'll find it growing completely underwater in freshwater ponds, lakes, sluggish streams, and canals. Look for coontail anytime during the summer in just about any quiet patch of freshwater, ranging from wilderness lakes to ponds in city parks. You'll usually see coontail growing in moderately shallow water near the shore.

Coontail plants have long, skinny stems and narrow little leaves. The stems will usually have scattered side branches. Some of the coontail plants you'll see may be only four inches long, while others may be nine feet long. The olive-green leaves grow in whorls, several leaves spreading out from around the same node in the stem. In the center of the plant the whorls of leaves may be a couple of inches apart. The whorls of leaves grow much closer together near the end of a branch. This gives the tips a dense, bushy, raccoon-tail-like appearance, from which coontail gets its name.

The narrow leaves are only ½ inch to ¾ inch long, so a coontail plant is very long and narrow. The key to identifying the plant for the first time is to pluck a leaf and look closely at it. Coontail leaves fork and have tiny hornlike projections. The horns may be very small, but they've given rise to another common name for coontail: hornwort.

Break a branch off a coontail plant and you'll have two coontail plants—the branch will float free and keep growing. In late summer some coontail stem tips grow so compact that they break off and sink in the water. Next year's plants grow from these tips, which survive the winter embedded partway in the bottom of the pond. Coontail never grows roots, but part of the stem will usually stay embedded in the bottom of the pond. In spring the small coontail plants grow straight up until they reach the water's surface and then spread out sideways, with the main stem branching repeatedly. By midsummer you'll often find a tangled mass of coontail and other aquatic plants floating just below the surface of the water.

Boaters don't appreciate mats of coontail. The dense growth slows canoes and entangles the propellers of motorboats. Swimmers, too, don't like the interference or the feel of coontail. Wildlife appreciate coontail, though. Muskrats and some ducks eat coontail, especially coontail seeds, and small fish and insect larvae use coontail as shelter from predators.

Duckweed

Duckweed is a small, free-floating plant that grows in ponds, lakes, rivers, streams, swamps, and even wet ditches. Duckweed floats to wherever winds or currents push it, so the best places to look for duckweed are sheltered areas such as a swamp or near the shore of a pond. By midsummer the tiny duckweed plants may have grown so well in these places that they form a solid carpet of green over the water.

Duckweed doesn't have any stems, and it doesn't have any leaves, either. What looks like a tiny leaf is actually a small bit of multipurpose plant tissue called a frond. Duckweed fronds are thin, like leaves, oval or circular, and only about ⅛ inch long. Dangling down from the floating frond is a hairlike rootlet, which may grow as long as a couple of inches. That is all there is to one complete duckweed plant.

Duckweed is one of our smallest flowering plants. With the plants only ⅛ inch across, imagine how small the flowers are! Duckweed flowers are very hard to find, not only because you'd need a magnifying glass to see them, but also because they are rare. Most duckweed reproduction comes from one plant growing an extra frond or two until the extra fronds finally break off to form their own plants. This reproductive process is so continuous that most of the duckweed plants you'll see consist of two, three, or four fronds still stuck together. And if the water has plenty of nutrients in it, this growth process may go on very fast. Some ducks and other waterfowl as well as certain fish eat the prolific duckweed.

Duckweed is one of three major groups of small, oval, free-floating plants: big or great duckweed, regular duckweed, and watermeal. They are very easy to tell apart. Big duckweed (A) is a little larger than regular duckweed and has several rootlets attached to each frond. Regular duckweed (B) has one rootlet coming down from each frond. The undersides of big duckweed fronds also tend to be a little purplish. Watermeal (C) is smaller than regular duckweed and does not have any rootlets at all. A watermeal plant is frequently only the size of a pinhead.

Waterlilies and the
American Lotus

In the shallow part of a lake or backwater of a river you will find many types of plants growing right up through the water. Among the most interesting of these are the plants with huge, floating leaves. If at least one leaf is about ten inches or more across, then the plant can only be a waterlily or a lotus. You can easily tell which one it is. Think of the leaf as a pie. If the pie is whole, the plant is an American lotus. If a piece is missing from the pie, the plant is a waterlily, even if the leaf edges overlap where the gap is. In the early spring, before waterlily and lotus leaves have grown to their distinctively large size, they are hard to distinguish from other smaller-leafed plants such as floating heart and water shield.

We have several different species of waterlily in North America, and each of their leaves has one notch that goes all the way to the stem. Thanks to the notch, the stem can come into the leaf at an angle, allowing the leaf to float flat on the water a short distance from the center of the plant. Come midsummer, waterlilies will send up separate stems with buds that open into large, beautiful, and sometimes wonderfully fragrant flowers of various colors.

There is only one species of American lotus, usually found in the eastern half of the United States. Its leaves are nearly circular, with its stems attached in the center, which often makes the top of the leaves dip down a little bit at that point. Most of the time you'll find lotus leaves floating on the water, but sometimes the stems will push the leaves clear out of the water, making them hang in the air like oddly shaped umbrellas. The American lotus blooms in midsummer to late summer, with large pale yellow flowers. The central part of the flower eventually turns into a salt-shaker-like seedpod, with each of the large seeds locked in its own open-topped socket. Both seeds and roots of the American lotus are eaten by wildlife, especially muskrats.

American lotuses are related to the lotuses that were considered sacred in Egypt and India. One reason lotuses were held in such high esteem was that they set an example for how we should all live, for they produce flowers of great purity and beauty despite growing in the muck and mire of swamps.

Rushes and Bulrushes

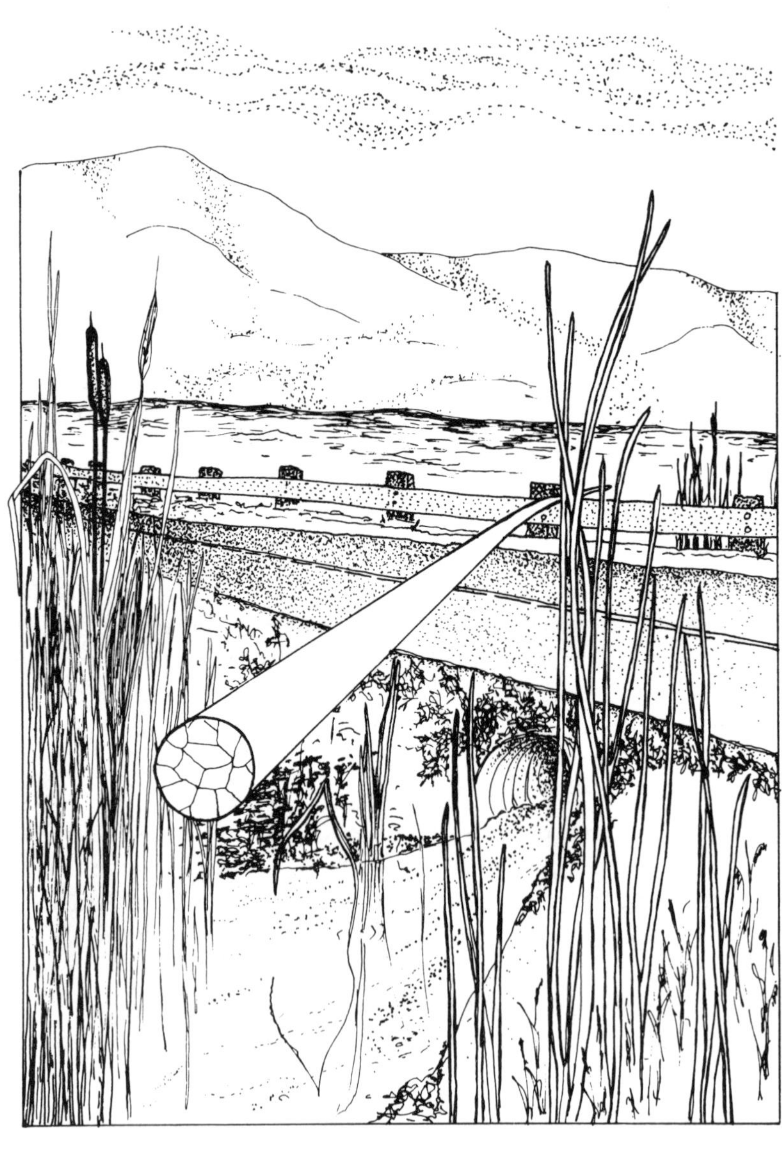

Rushes and bulrushes appear to be all stem and no leaves. The round, pointed, green stems of rushes and bulrushes do the work of leaves by carrying on photosynthesis. Most rushes and bulrushes do have a few leaves, but not enough to interfere with their identification: Any grasslike plant made up almost entirely of round, pointed stems is a rush or a bulrush.

Marshes are the best places to find rushes and bulrushes. These plants grow well in shallow water. They spread by roots as well as by seeds, so you'll usually see them growing in clumps or dense stands. You can also find rushes and bulrushes growing in all sorts of places where the soil is fairly moist, so look for them in ditches along the sides of roads, in moist meadows, and even in some lawns.

Many species of rushes and bulrushes grow in North America, and they come in all sizes. A rush species found in a lawn might grow only a foot tall if allowed to reach its full height, while certain species of bulrushes reach nine or ten feet in a fertile marsh.

Some species of rushes and bulrushes are practically identical, yet rushes and bulrushes aren't related. Rushes are related to lilies, and bulrushes are related to sedges. So how do you tell rushes and bulrushes apart? The answer is, you don't. Botanists can tell them apart by differences in their drab little flowers, located near the pointed ends of the stems, and present for only part of the summer, but to most of us they are virtually the same. Consequently, rushes and bulrushes are always being mixed up, with individual species being called a rush in one part of the country and a bulrush in another part.

Rushes and bulrushes provide food and shelter for many creatures. Muskrats eat the plants. Mallard ducks and other waterfowl eat the seeds off the bottom of the marsh. People, too, come to marshes to harvest rush and bulrush stems, which are used to make thatch roofs and woven chair seats. In the days before carpets they were strewn loose across cold stone or dirt floors. So to get a sense of history, don't feel the round stems with your fingers, feel them with your toes!

Sedges

Grasses have round, hollow stems, so when you find a grasslike plant with a triangular stem, you know right away that it isn't a grass. It may grow like grass, and it may have long, pointed leaves like grass, but the triangular stem tells you it is not a grass; it is a member of the sedge family. Just remember "grasses grow round, sedges have edges."

The best place to find one of the many North American species of sedge or nut sedge is a moist meadow near the edge of a marsh. If the meadow has one or two dominant species of grass, look closely at the grass color. Then scan the meadow and investigate patches that have slightly different shades of green. You can also find sedges right in marshes, along shores of lakes and rivers, and in moist meadows far from open water. Nut sedges can also be found in farm fields and even gardens.

Not all members of the sedge family have triangular stems, but all the grasslike plants you'll find with triangular stems are sedges or nut sedges. The major difference between sedges and nut sedges is found underground, in their roots. Sedges spread underground by growing large, horizontal, rootlike rhizomes that sprout new plants. Nut sedges send out small, horizontal roots that grow little lumps, or "nuts," underground. New nut sedge plants will grow up from these little lumps.

The triangular, pithy stems of sedges and nut sedges have three flat sides that meet at fairly sharp angles. Sedges' grasslike leaves have creases down their centers that fit nicely over any corner of the stem. The triangular stem provides stiff support for the leaves and seed head. Such a triangular configuration is very rare in nature. Most animals can be divided evenly into symmetrical halves, and most plants are organized around a circular stem. If the sedge's triangular stem had major advantages over round stems, you'd think more plants would have triangular stems.

Scouring Rush

Scouring rush's name holds the two keys to its easy identification. Scouring rush is not a rush, but a member of the horsetail family (see page 32). However, it does look very much like a rush. The plant has no leaves, just long, round, green stems. Unlike rushes, however, scouring rush stems are jointed at regular intervals.

Scouring rush also got its name from being used for scouring pots and pans. The plant makes a good scouring pad because its rough and tough stem is made up of vertical grooves and abrasive silica crystals. Silica is a major ingredient of sand, clay, granite, quartz, and glass. If you look closely at a scouring rush section in bright sunlight, you can see the silica crystals as minute sparkles. With a hand magnifier or microscope, the sparkles resolve into little clear lumps on scouring rush's ridges. Test the roughness of the stem with your fingers when you're trying to identify scouring rush.

You can find this plant growing on banks and gravelly roadsides, in moist woods and grasslands. It is typically from ¼ inch to a little over ½ inch thick and from a foot to 4 feet tall. If you are looking for it in moist grassy areas or along roadsides, late spring to early summer is the easiest time to find it. In late spring scouring rush is usually taller than the surrounding grass. By midsummer the grass has caught up and road crews may have mowed the roadsides. In moist woodlands you'll probably find scouring rush a little easier to locate in the fall, as the leaves are dropping. If the ground is moist, scouring rush will be there all year round, so keep your eyes open.

Scouring rush spreads by spores and by creeping roots, so where you find one scouring rush you'll probably find several, perhaps even a dense stand of them. If the plants are growing well, pick one. You'll notice that the stem is hollow between the joints, and that the joints snap apart. Besides cleaning dishes, scouring rush has been used to sand wood and polish metal. Take the sections home and try scouring or sanding with them. You may be surprised at how nicely this primitive tool works.

Horsetail

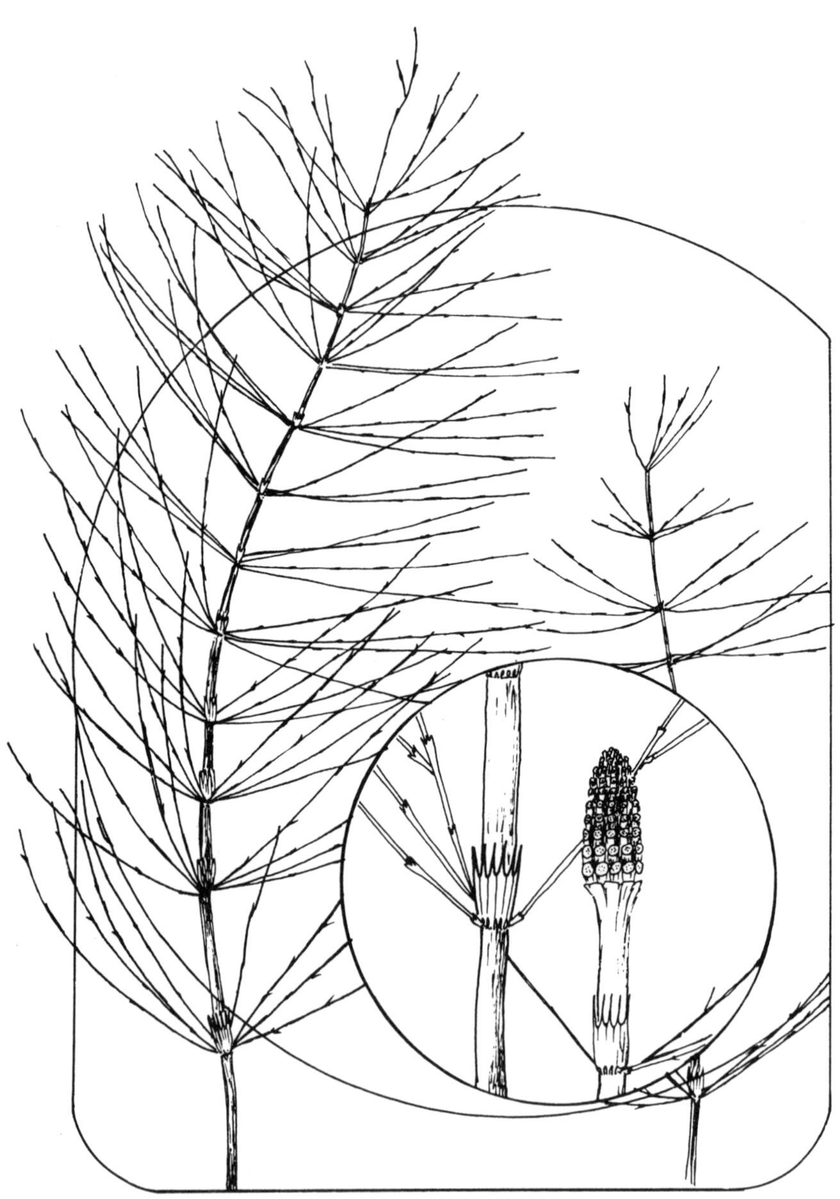

Horsetails are relics of Earth's primeval past. Long before horses evolved, Earth had huge forests of gigantic plants very much like these, with a weedy stem that grows one section on top of another, with each joint growing a whorl of pine-needle-like "leaves." And even the pine-needle-like "leaves" grow in sections, with one section attached to the end of the section before. Technically, the pine-needle-like "leaves" are actually branches. But they are green and carry on photosynthesis like leaves. The name horsetail comes from the plant's resemblance to horses' bushy tails, but I sometimes think they look more like miniature pine trees.

Most of North America's modern-day horsetails only grow half a foot to three feet tall, but the giant horsetail of the Pacific Coast sometimes reaches six feet. The easiest place to find them growing is in the gravelly soil at the sides of roads. Horsetails grow well in wet sandy or gravelly soil. You can also find horsetails growing in forests, moist fields and meadows, and marshes and swamps.

Look for horsetails anytime during the summer. The shoots that look like horses' tails or miniature pine trees come up in late spring and normally last until fall frost. Horsetails are too primitive to have flowers. They spread by roots and spores. Spores serve much the same role as seeds, but spores are created without sex. The most common horsetails send up special little stalks in early spring that don't have the long, leaflike structures. These stalks make the spores, shed the spores, and quickly wither away.

Like scouring rush (see page 30), horsetails feel rough because of silica crystals in their stems and "leaves." You'll notice that the early-spring horsetail shoots look much like short scouring rushes. Unlike scouring rush, though, the spring shoots of horsetails have collars around the joints that look as if they might keep growing out into "leaves."

You may be surprised that a plant innocently named after part of a horse would be poisonous to horses, but it is. Horses or cattle that eat large quantities of horsetails get sick and sometimes die.

Puffball Mushrooms

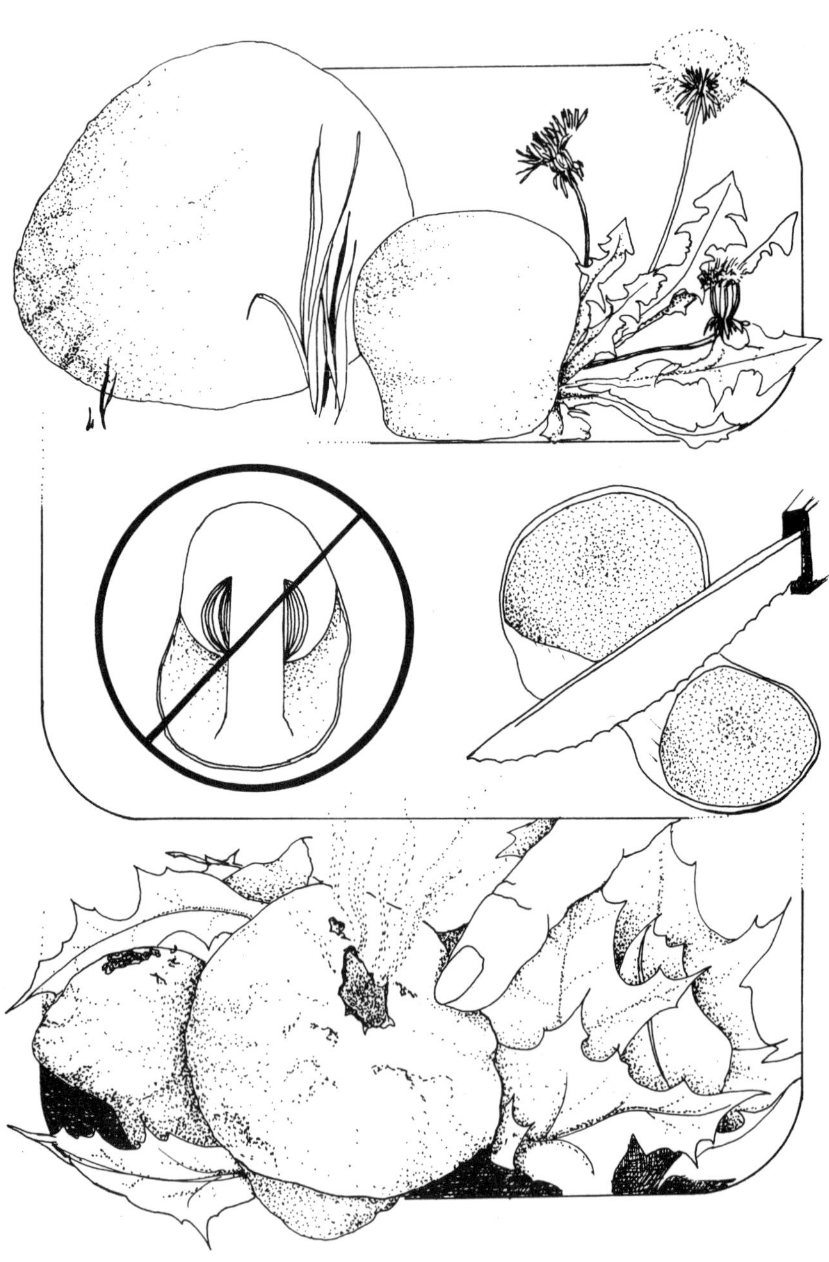

The puffball mushroom's name gives you the key to identifying mature puffball mushrooms (young puffballs do not puff). They are shaped somewhat like balls, and puff out a smoky cloud of powdery spores when you tap the sides of the ball. Puffball mushrooms vary greatly in size. Some grow as large as soccer balls, while others are only the size of marbles when fully grown.

Young puffballs are usually light-colored and vary from smooth to quite rough and warty. You can recognize puffballs because they are roundish or pear-shaped and lack true stems. But to positively identify young puffballs, you will have to slice the mushroom in half from top to bottom. Regular mushrooms go through a stage where they are shaped much like puffball mushrooms. When you slice open a regular mushroom at that stage, you will see evidence of the growing cap with a stem in the middle. But when you slice open a young puffball, you'll find that the entire inside is a soft, uniform mass that is usually white. Sometimes only the top part of the inside forms spores, which gives a two-toned coloring to some older puffballs. As the spores ripen, the inside turns powdery and darker. A large, mature puffball may hold trillions of powdery spores.

A hole usually forms on the top of the puffball to allow the mature spores to escape in little puffs when raindrops or animals hit the mushroom. At this stage the outside of the puffball will typically be gray or brownish and look like a partly deflated ball. Some puffballs never get a little hole on top. Instead, the outside of these puffballs just falls apart.

You'll often find puffballs growing in lawns and city parks as well as in meadows, forests, and deserts. Puffballs are most common from summer to late fall. Look for them on the ground a couple of days after a heavy rain. There are many different species of puffball mushroom, all with the same characteristics, and they can be found in many different habitats and at different times of year. So keep your eyes open for puffballs in the spring, too.

To find giant puffballs in east and central North America, look in wet meadows and at the edge of woods in late summer or early fall. The western giant puffball mushroom is more common in spring and summer. In western North America, look for giant puffballs under sagebrush, in prairie grass, and in wet meadows.

Spruce and Pine Trees

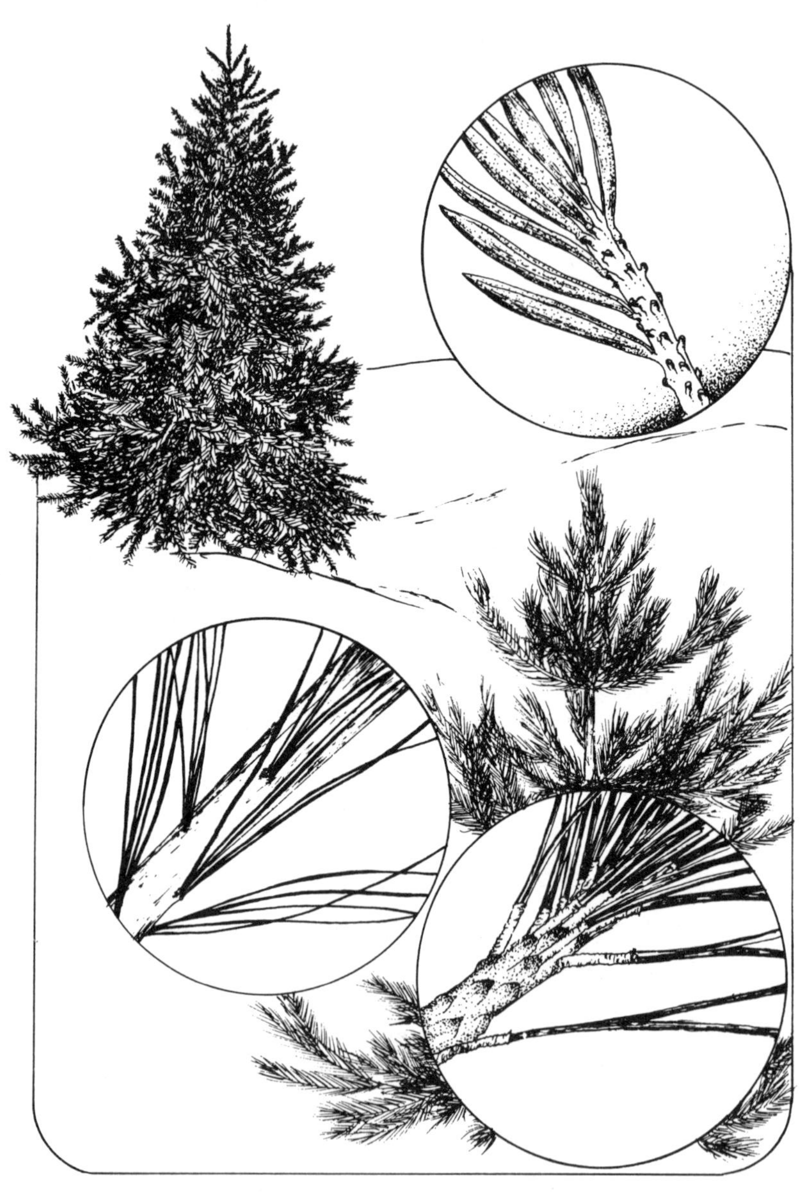

Spruces and pines are both conifers, and are easily distinguished from the many others, such as cedars, firs, larches, hemlocks, and junipers. You identify spruce trees by the short, woody, peglike bases from which the individual spruce needles grow. Look closely at a twig where some needles have recently fallen off or pull off a couple of needles yourself to see the curved projections.

Spruces are common in the far north and can also be found growing wild in mountains as far south as Arizona and North Carolina. The easiest places to see spruce trees are yards and parks in towns, where the straight, tall, sharply pointed spruce trees are commonly planted.

You identify pine trees by their needles, growing in bundles or groups of *two to five* needles arising from the same place. The sole exception to this rule is one of the southwestern piñon pines, on which the needles grow singly. All young pine needles, and in some species, old needles, have a papery sheath growing around the base of the bundle.

You can tell about how old many pine trees are by counting the sections of trunk between sets of branches and estimating where branches have fallen off the bottom part of the trunk. Every year a pine will normally grow one new set of branches and send its leading stem up one space. The distance the leading stem grows reflects how good a growing season the pine had the year before. Counting back the years on a pine can tell you which years were good growing years and which were poor, perhaps because of drought.

How do pines stay evergreen when they shed enough needles to carpet the ground below? Like deciduous trees, every spring pine trees grow new leaves and every fall they shed old leaves. But pines don't shed the old leaves until the needles are between a year and a half and several years old, depending on the species, so they always keep their color.

In the Northeast you can easily identify the eastern white pine. It is the only wild or commonly planted pine in the Northeast that has five needles per bunch.

Pines can be found growing wild in many parts of North America, but the easiest ones to find are those that have been planted in towns.

Bloodroot

A bloodroot plant appears to consist of only one white flower accompanied by one leaf. The flower is about 1¾ inches in diameter, which is quite large for a plant that grows only 3 to 6 inches tall. Bloodroot's name holds the key to its positive identification. To identify bloodroot for the first time, dig with your fingers down an inch or two into the earth under a low-growing white wildflower and expose part of the root. Before replanting it, nick the root with your thumbnail. If the inside is orange-red and oozes a bloody, orange-red sap, the flower is bloodroot.

You'll find bloodroot in the eastern half of North America from Nova Scotia to Manitoba and south to Texas and Florida. Look for bloodroot on sunny days in spring, before the leaves of trees have gotten very large. Bloodroot is generally considered a woodland wildflower, but it can also be found in moist meadows and along streams. The best place to look for bloodroot is near the edge of a deciduous woods. There bloodroot sometimes grows so thick that the blossoms almost carpet the ground.

Indians used the fat bloodroot roots, which are actually horizontal underground stems called rhizomes, for a dye. One rhizome will send up many sets of single flowers with single leaves, so bloodroot plants are much larger than they appear to be from above ground.

You'll need to see a bloodroot rhizome only once. After it has told you positively which wildflower is bloodroot, you'll be able to recognize bloodroot by the leaves and flowers. Look closely at bloodroot and you'll notice that the leaves are deeply lobed and grow sort of wrapped around the flower stem. The flowers are a little taller than the leaves and have eight to twelve petals. Often the petals aren't all the same size.

Bloodroot blooms early enough in the spring to get most of its growth in before tree leaves shade it too much. Growing this early in the year has its difficulties. One problem is attracting bees to the flowers when the weather is cold. Some researchers feel that the bright white petals of bloodroot and some other early wildflowers act as solar reflectors to help warm the bees when they are on the flowers. Another problem is protection from bad weather. Growing very close to the ground offers some protection. Also, bloodroot flowers close up during inclement weather, so look for bloodroot on sunny days.

Jack-in-the-pulpit

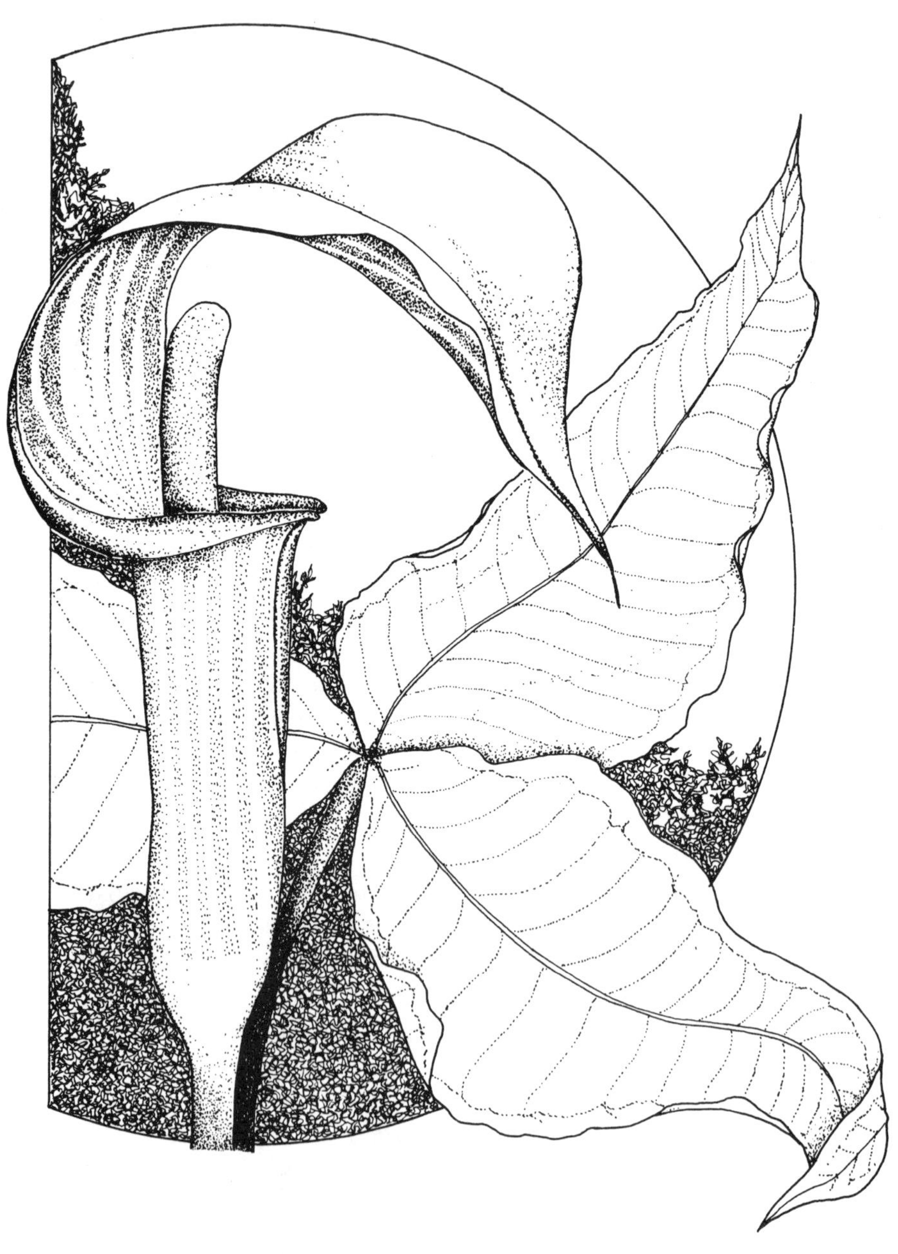

Jack-in-the-pulpit is such an unusual wildflower that you'll have no trouble identifying it when you see it. However, jack-in-the-pulpit doesn't have any bright colors, so you'll have to actively look for it. Look for this greenish wildflower in late spring. You'll find jack-in-the-pulpit growing in forests and swamps. The best places to look for jack-in-the-pulpit are wet woods or north slopes of wooded hills. The sun is less direct on a north slope, so the ground is moister. Jack-in-the-pulpit is an eastern wildflower. It can be found from Nova Scotia to Florida, but only as far west as Manitoba and eastern Texas.

The key to identifying jack-in-the-pulpit is "Jack" in his "pulpit." "Jack" is a club-shaped growth several inches long, standing straight up. His "pulpit" is a leaf-type growth that surrounds the bottom three quarters of "Jack" like a vase. The top of the "pulpit" curves up in the back to form a roof over "Jack's" head.

A jack-in-the-pulpit plant has only one or two leaves. Each leaf, though, has three segments, or leaflets, that look like separate leaves. The leaflets are oval-shaped with points at the ends. A full-grown jack-in-the-pulpit plant varies from one to three feet tall and ranges in color from light green with dark green markings to dark green with purple streaks. These colors apply to "Jack" and his "pulpit" as well as the leaves and stems.

Inside the "pulpit," sheltered from wind and rain, grow jack-in-the-pulpit's flowers. "Jack" is a flowering spike, though the part you normally see of "Jack" doesn't have any flowers. The tiny flowers grow only on the lower part. Flip up the "pulpit's" roof and you'll probably see some tiny bugs inside the "pulpit," busy fertilizing the flowers.

Where "Jack" and his "pulpit" stand in the spring, a tight bunch of bright red berries will stand in late summer. Jack-in-the-pulpit leaves die back once the berries, which contain the seeds of new plants, have formed and enough energy for next spring has been stored in the plant's bulb.

If you have a moist, shady place, such as the north side of a house, where you'd like to grow this unusual wildflower, you can buy jack-in-the-pulpit plants at some nurseries. If your local nursery doesn't sell them, look through the mail-order nursery catalogs available at many public libraries.

Mayapple

Mayapples are woodland wildflowers not related at all to regular apple trees. You'll frequently find mayapples flowering in May, with little green applelike fruit following later in the summer. Each plant produces one plum-size "apple" (not edible), which ripens to lemon yellow in July or August.

Mayapples grow in the eastern half of North America from Quebec to Florida and west to Manitoba and Texas. You'll find mayapples growing in all sorts of woods and occasionally in the open. They grow best in partial shade, so the best places to look for mayapples are woods with some light reaching the forest floor. Mayapples are quite interesting to see from late spring, when they are flowering, through midsummer to late summer, when they are fruiting.

The key to mayapple identification is its umbrellalike leaves. The leaves are one or two feet off the ground and may reach almost a foot in diameter. The stem is attached near the center of the roundish leaves, and the leaves stick out like an open umbrella. Examine the leaves and you'll notice that instead of being solid, mayapple leaves have many deep divisions in them, which give the leaves five to nine lobes. The large leaves are the top of mayapple plants. Below the leaves will be either a straight stem going to the ground or a Y-shaped stem connecting two leaves to the ground. The stems with two leaves have a single flower attached at the junction of the leaf stems.

Mayapples generally bloom during April and May in the South and during May and June in the North. You'll have to stoop over and peer under the leaves to see the one- to two-inch-diameter white flowers. A mayapple's flower, and later its fruit, is attached to the end of a short, arching stalk.

You'll often see mayapples growing in colonies, because mayapples spread through their rootlike underground stems as well as by seeds. Frequently these mayapple colonies grow so thick that the leaves touch and prevent you from seeing the ground for many yards around. All the mayapple leaves in a colony are usually about the same height and appear to form a carpet floating one or two feet off the ground.

Wooded pastures often have colonies of mayapples, because cows leave them pretty much alone. And for good reason—mayapple roots, stems, leaves, flowers, seeds, and unripe fruit are poisonous.

Wild Columbine

With five colorful red-and-gold hollow prongs curving gracefully upward, columbine flowers look like five elves' hats. Indeed, the delicate columbine doesn't appear tough enough to compete in the real world of fields and forests. And to some extent, columbine can't compete head to head with grasses and broad-leafed weeds. In a fertile field the leggy columbine would be crowded out by grasses. But columbine can compete and grow where grasses find life more difficult, such as in rocky places and open woodlands. Columbine is easiest to find if you look for it in places that combine both of these features. A rocky outcropping that is in woodland but still receiving some sunlight would be ideal.

Columbine plants are usually from one to three feet tall and present a loose appearance. They have a lot of branching stems but are not densely leafed. The best time to look for columbine flowers is in early summer, for all the columbine species bloom then. But since some columbines start flowering in April and at least one species is still flowering in September, you may want to keep your eyes open for them at other times as well.

The key to columbine identification is the five beautiful flower petals, the elves' hats, that point back away from the flower. These petals, called spurs, are open at the flower end and closed, sometimes with a little ball, at the other end. Hummingbirds often find a drop of nectar at the closed end of the spurs. All five of the hatlike petals are joined together in the middle, with the stamens and pistils extending beyond the open end of the hats. The average columbine flower is about 1½ inches long and droops over so it is hanging downward with the spurs pointing up.

In the eastern half of North America all the columbine flowers you'll find will be red on top and yellow on the inside and underside. In the West some species of columbines are red and yellow, but other species are almost all yellow, and one is blue. The spurs on some western columbines are very narrow, and the flowers of one species grow upright rather than droop. But every single columbine flower will have five distinctive petals shaped like hollow prongs projecting backward from the flower.

Stinging Nettle

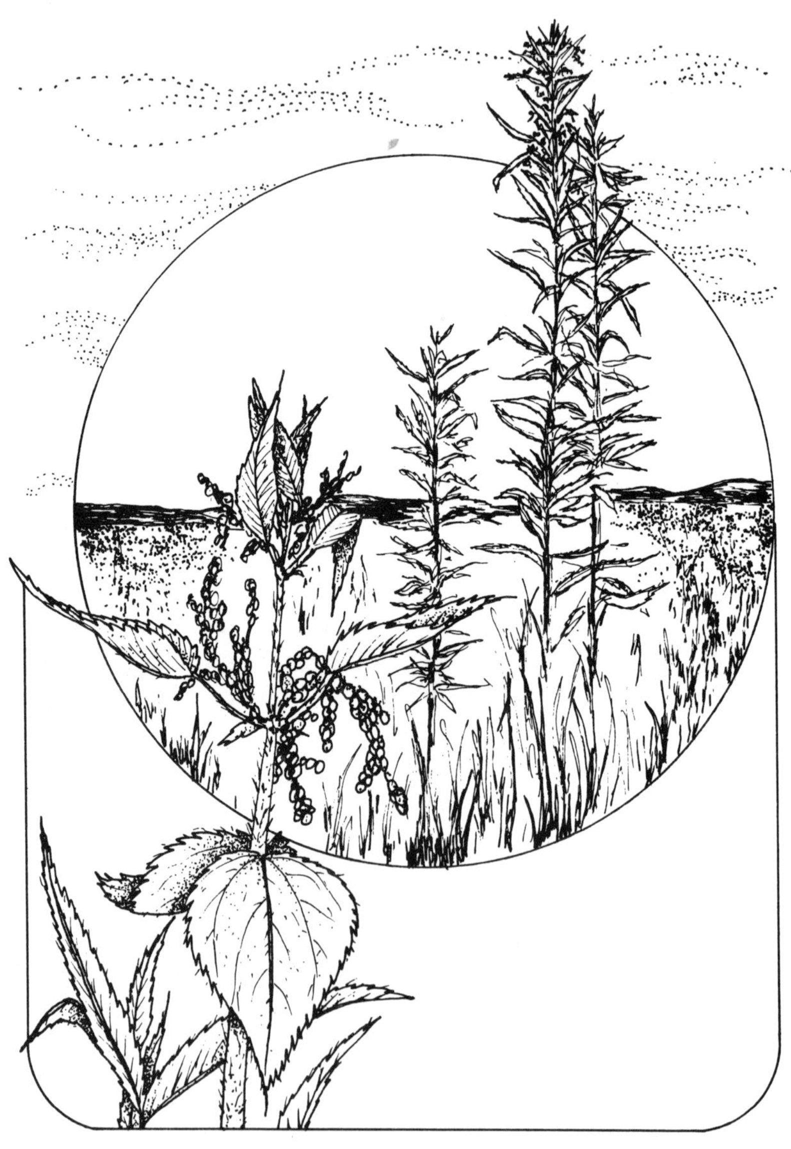

The first time most people identify stinging nettle is right after they have been stung. Until then, stinging nettle was just one more green weed. After they are stung, however, stinging nettle's toothy green leaves look distinctive.

The keys to identifying stinging nettle are long, erect stems; bumpy-edged leaves that are opposite, meaning that where one leaf grows, there will be another leaf growing from the same place but on the opposite side of the stem; and stinging hairs sticking out from the sides of the stems, leafstalks, and sometimes leaves as well. This identification won't enable you to look over a field and spot stinging nettle if you've never seen it before. But when you are out for a walk and get stung by stinging nettle, you will be able to turn around and identify the culprit. This will help you avoid it in the future.

You can run into nettles just about anywhere, from stream banks in cool woods to vacant lots, roadsides, and neglected yards. Stinging nettle starts growing very early in the spring and doesn't die until fall.

Most of the stinging nettle you'll see will be from two to six feet tall, but stinging nettle can be as small as six inches or as tall as ten or twelve feet. This variation in mature height is due mostly to how fertile the soil is. You'll often find nettles growing in rich soil, where they can grow larger and compete more effectively. Some of the variation in size is also due to the fact that North America has six different species of stinging nettle.

A typical nettle sting feels like a mild beesting and goes away completely in twenty minutes. With six species of nettle, not all stings will be typical. Some may just cause intense itching, and some will last longer.

Nettles spread by creeping roots as well as seeds, which are eaten by winter songbirds such as chickadees and goldfinches. So where there is one nettle plant, you'll often find many. When you find a patch of stinging nettle growing lush and tall in a cow pasture that has most of the grass and other weeds eaten down, you'll realize how effective the stinging hairs are.

Velvetleaf

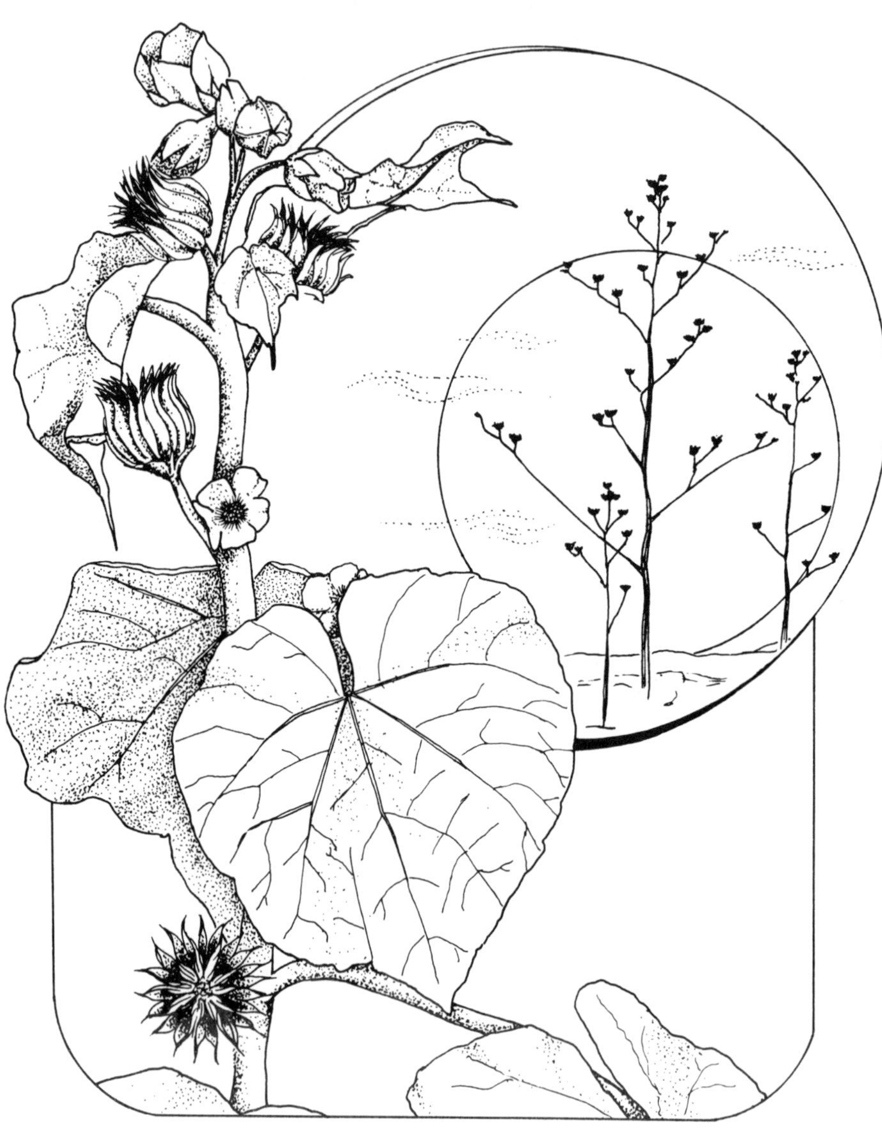

Velvetleaf is a weed that dies completely every fall and must grow from seeds in the spring. The seeds and young plants find it easiest to grow where the ground has been disturbed in some manner that would hurt the root system of perennial weeds and grasses, so look for velvetleaf in gardens and cultivated farm fields as well as in empty lots in towns and other waste places, anywhere the ground has been disturbed.

The key to identifying velvetleaf is the soft, slightly fuzzy, heart-shaped leaves with the feel of fine velvet. You'll have to touch and feel these unusually soft leaves to be sure. You'll also, of course, have to generally know what the plant looks like so you'll know which plants to give the final touch test to.

Velvetleaf plants seem to grow in units of one. They usually have one central stem attached to one straight taproot, and the leaves come off the stem one at a time on long petioles, or leafstalks. A full-grown velvetleaf plant may be as short as one foot if it is growing on infertile ground or as high as seven feet if growing in a fertile cornfield. Most of the velvetleaf you'll find will probably be three or four feet tall.

Velvetleaf seems to grow particularly well in corn and soybean fields, so corn and soybean farmers look on velvetleaf as a bothersome weed. Velvetleaf, however, hasn't always been looked upon as a weed. This plant was originally imported from India to be grown as an ornamental. During the summer, when you see and feel a soft, green velvetleaf plant, you'll know why. Even during the winter, when velvetleaf is dry and brown, it is still interesting to look at. You might even want to pick some velvetleaf in winter to give an Oriental flare to a dried flower arrangement.

As a wildflower, velvetleaf is not particularly showy. A velvetleaf plant usually grows several one-inch yellow flowers between July and October. The flowers give way to interesting seedpods, each with a ring of prickly horns curving out from around the top. These seedpods have traditionally been used to make fancy marks and patterns on piecrust or butter. To continue this old tradition, you may want to save a couple of the seedpods for the next time you have an opportunity to get involved in pie making.

Chicory

Chicory's blue flowers growing on nearly bare stems are real eye-catchers. This is true partly because of their own beauty, and partly because blue wildflowers are not as common as yellow wildflowers. Chicory flowers stand out even more because they open up flat like a daisy, while most of our blue wildflowers, such as bluebells or wild iris, tend to be funnel- or tube-shaped. There are other flat blue wildflowers, though, so you will need to follow a three-step process to positively identify chicory.

You can be sure the plant is chicory if: (1) The flowers are daisylike with blue petals and about one inch wide or slightly wider. (2) Some of the flowers are attached directly to the sides of major stems. The blossoms of most wildflowers grow only on flower stalks. Some chicory flowers grow on short flower stalks, but others are attached directly to major stems. And (3) the main stems are rough and nearly leafless. Most of chicory's leaves grow next to the ground. A few leaves grow higher up on the stem, but these are small and let a lot of bare stem show through. If you feel the stem with your fingers, you'll notice that the stem is coarse. Most chicory grows from one to four feet tall.

Chicory is one of the most urban of our wildflowers. You can find it growing in empty lots, in alleys, and along roadsides. Chicory survives in urban areas because it can tolerate being mowed occasionally. Also, it has a strong taproot that enables it to grow in poor, gravelly soil. This is not to say that chicory isn't found in the country. Quite the contrary—chicory is found on the shoulders of rural roads and sometimes becomes common enough in farm pastures to be a problem weed. Chicory can grow taller in fertile soil, but there it tends to get shaded out by grasses and weeds that have leaves growing up high.

The best time to find chicory blooming is from June to October. Chicory is a European import that has spread far and wide in North America, but it is not very common in the Southeast.

Queen Anne's Lace

A typical Queen Anne's lace flower looks like an intricately woven white lace medallion three to four inches in diameter. Each floral medallion is a combination of many little white flowers growing on perhaps fifty separate little stems. Each stem is a different length, allowing the entire array to appear quite flat. One small, dark, brownish-red floret usually grows right in the center of the white Queen Anne's lace flower, almost like the jeweled head of a pin that might have fastened a lace medallion on Queen Anne's dress.

When identifying Queen Anne's lace for the first time, look under a flower at the small, green, leaflet-type things growing just below the flower stems. If these bracts, as they are called, divide evenly along a center axis into sharp, narrow segments, then the flower is definitely Queen Anne's lace. Once you've gotten to know Queen Anne's lace, you'll be able to recognize it at a glance.

You'll find Queen Anne's lace commonly growing along roadsides, in waste places, and in pastures. The plant grows one to five feet tall and may be blooming anytime from May to October. Motorists seeing tall, many-flowered Queen Anne's lace along a roadside usually consider it a wildflower. Farmers seeing rough, bristly stemmed Queen Anne's lace growing in a pasture usually consider it a weed.

Queen Anne's lace is a wild relative of the carrot. You can notice its similarity to carrots by looking at the leaves. Both Queen Anne's lace and carrot leaves have so many divisions in them that the final segments are quite small and narrow. Queen Anne's lace and carrots are also both biennials, which means that during the first year of growth they store a lot of energy in the root. During the second year they use this energy to produce flowers and seeds. When Queen Anne's lace goes to seed the flowers dry and curl up, forming a floral "bird's nest," or basket.

Prickly Pear Cactus

Prickly pear cactus is common in the southwestern desert. It is also fairly easy to find in moderately dry areas throughout the West. The several species of prickly pear cactus are not, however, limited to the arid West. You can find isolated patches of prickly pear growing in quick-draining sandy or rocky soil even in areas of heavy rainfall.

Prickly pears are easy to identify because they are made up almost entirely of flat, prickly, pear-shaped sections of stem. Each green section of stem serves the same function as leaves but is much, much thicker. A new flattish, pear-shaped section of stem will grow off the top of an old one. Prickly pear plants look as if they would like to grow tall but the stems aren't strong enough, so the plants tend to tip over and sprawl along the ground. Some prickly pear plants reach six feet in diameter and may grow several feet tall.

Prickly pear flowers are usually bright yellow. Look for them from March to May in the southern part of the United States and from May to July in the northern United States and southern Canada. A prickly pear in full bloom is a sight to remember. It is loaded with large, soft, alluring flowers, a vivid contrast with the rest of the plant, which warns people away with either spines or barbed hairs. Prickly pear spines can be long and sharp, but the barbed hairs are said to be more dangerous because they can detach and work their way into flesh, causing a painful inflammation.

If you live in humid areas as far north as Ontario or as far east as Massachusetts or Georgia, you may want to look for prickly pear on rock bluffs or sand dunes. Or, if you happen to be out hiking and see a prickly pear cactus growing, you will know that the soil in that spot is sandy or rocky, and that rainfall must drain quickly through the soil.

Arid conditions or soils that dry out quickly after a rainstorm are attractive to prickly pear because prickly pear's competitors, such as grasses, do not grow very well under such conditions. You may find grasses and weeds growing there, but they will not be as strong or robust as they would be with more moisture. Prickly pear can tolerate arid conditions well because it doesn't have any leaves to lose moisture from and it can store water in its thick-skinned sections.

Common Mullein

Common mullein (pronounced MULL-in) is called flannelleaf in some parts of the world. The reason for that name is also the key to common mullein identification: Both sides of common mullein leaves are covered with a thick layer of fuzz that makes them feel like soft flannel.

Common mullein is a biennial that produces a circular cluster of oblong leaves the first year. These leaves usually manage to stay soft and green right on through the winter, even under snow. The second year mullein grows a stem. And what a stem: straight, tough, and usually four or five feet tall. The leafy stem ends with a whole series of little yellow flowers. The little flowers aren't very showy, because they flower only a few at a time from June to September.

Called both a wildflower and a weed, common mullein can be found along roadsides and fencerows and in uninhabited or neglected areas like vacant lots, weed patches, etc. Common mullein's root system penetrates quite deeply, allowing it to grow in fairly dry and sterile soil that doesn't support many other big weeds. You can also find common mullein in pastures, because most livestock don't particularly like eating it. The thick fuzz on the leaves, which feels so soft to the gentle touch of fingers or toes, apparently discourages some animals. To understand why, take a look at mullein fuzz through a magnifying glass or microscope. You'll be surprised to see how individual mullein hairs branch off, like miniature leafless trees, with each branch ending in a sharp point. And the fuzz is nearly everywhere—on both sides of the leaves, on the stem, and even a little bit on the flower petals. The sharply pointed fuzz can feel soft to the fingers yet still poke and irritate the inside of an animal's mouth.

Mullein grows throughout most of the country. How large depends a great deal on such factors as soil fertility and competition for water. Sometimes the leaves in the first year's rosette will be less than six inches long. Other times the leaves will be eighteen inches long. The second year's stem may only be a foot tall, or it may stretch well over six feet, with the top two feet reserved for flowers. Usually common mullein has only one flowering spike, but sometimes it grows several flower heads that branch off the main stem like a candelabrum.

Common Plantain

Botanists generally divide weeds into two groups. One group is the grasses and grasslike plants with sword-shaped leaves and parallel veins running down the length of the leaves. The other group consists of the rounder broad-leafed weeds with a network of veins branching throughout the leaves. The most striking feature of plantain is that it appears to fall into neither group. Its leaves are broadly oval, yet it has distinct veins running more or less parallel down the length of the leaves. Botanists place plantain in the broad-leafed weed group, though, and close inspection of plantain leaves reveals a network of veins spreading between the large parallel veins.

The key to identifying common plantain is the large, obvious veins running lengthwise down the entire length of the fat leaves. Common plantain leaves are broad ovals with a slight point on the tips. The leaves feel leathery and are quite thick and tough. All the leaves grow from the root, so plantain grows very low to the ground in a circular shape, with all the leaves radiating out from the center.

The easiest place to find plantain is in a lawn that isn't cared for too carefully. Manicured lawns treated with herbicide to keep out dandelions won't have plantain. You can also find plantain in many other places, such as paths, roadsides, and waste places. With all of plantain's leaves low to the ground, plantain can grow only where something keeps taller weeds from shading it out. Plantain's tough leaves survive foot traffic better than most grasses, so look for plantain in a part of a lawn where foot traffic has killed some of the grass.

Plantain is a long-lived perennial that sends up several narrow flowering spikes during the summer. Its pollen is carried by the wind, so plantain doesn't need to put on a showy display to attract bees. Consequently, plantain flowers are small and green. The flowering spikes will grow from several inches to a foot tall if they aren't mowed off. Plantain seeds are eaten and scattered about by birds.

Several different species of plantain grow in North America. Besides common plantain, with leaves at least half as wide as they are long, you may find a narrow-leafed plantain (also with fat veins) called buckhorn or English plantain.

Goldenrod

Goldenrod's name provides you with a good description of the plant. Goldenrod grows a single long, straight stem like a wand or a rod that is topped off with a large cluster of golden yellow flowers. You identify goldenrod by its tiny, ⅛-inch diameter individual flowers, which bloom at the very end of summer and in the early fall.

North America has almost a hundred different species of goldenrod, so you'll find that the shape of goldenrod leaves varies, but the leaves always grow singly. The shape of the flower cluster varies, but it is always at the top of the rod. The height varies from a few inches to seven feet, with three or four feet being common. Even the time of year goldenrods bloom varies somewhat, though most goldenrods start flowering at the end of summer. Many people have come to look on blooming goldenrod as a reliable sign that summer is almost over.

Various goldenrod species have adapted to life in deserts and marshes and many places in between. Look for goldenrod especially in pastures, old fields, at the edges of woods and along roadways. Goldenrod is a perennial that spreads by underground rootstock as well as by seeds, so you'll often find goldenrod growing in clumps.

Flowering late in the season has a distinct benefit: lots of bees. With fewer wildflowers around at the end of summer, bees really concentrate on goldenrod. You'll notice this right away when you walk through a blooming patch of goldenrod on a sunny early fall day. Goldenrod quickly forms tiny seeds with a little fuzz attached that helps the seeds drift in the wind. Birds eating goldenrod seeds in the winter also inevitably knock some seeds loose. Goldenrod dies down to the ground in the fall, and the dried stem stays standing through most of the winter.

Goldenrod often takes unfair blame for fall hay fever. Hay-fever-type allergies are triggered by windborne pollen. Goldenrod pollen is too heavy to float very well in the wind. It needs bees to carry the pollen from flower to flower, and it attracts bees with the showy yellow display. Ragweed often blooms at about the same time as goldenrod, only ragweed pollinates with airborne pollen. Since ragweed doesn't need bees, its flowers are plain and drab. A hay-fever sufferer sneezing as he looks at goldenrod may not even notice the ragweed plant flowering right next to it.

INSECTS

Goldenrod Galls

The best time to look for goldenrod stem galls is in fall and winter. These galls are about a ¾-inch–diameter swelling in a ⅛-inch–diameter stem. They were present in late summer when you were looking at goldenrod flowers, but are more noticeable after the leaves have dried and started to fall. If the patch of goldenrod is fairly large, it may have raised a large enough population of gall makers over the years to infest nearly every stem with a gall. Several different insects cause these abnormal swellings in goldenrod. Two of the most common stem galls are known as the ball gall and the elliptical gall.

The goldenrod ball gall maker is a little fly with brown markings on its wings. The fly lays its eggs on goldenrod stems in early summer. After an egg hatches, the maggot chews its way into the stem. In the center of the stem, the maggot continues to eat, but its damage is minor and the goldenrod stem grows larger around it. The ball gall maggot overwinters inside the gall. Then it pupates and turns into a fly in the spring. To see what ball gall flies look like, you can collect some galls in the fall, cut one open to make sure they have maggots in them, and then keep the galls in a jar with netting or tiny air holes on the top until the flies emerge in spring.

The goldenrod elliptical gall maker is a moth that lays its eggs on goldenrod in the fall. After an egg hatches in the spring, the tiny caterpillar climbs down the old stem, searches for a growing goldenrod stem, and chews its way in. Again, the insect does only minor damage and the goldenrod stem grows larger around it, but this time the gall is longer than it is wide. Before spinning a cocoon inside the gall, the caterpillar chews an exit hole and plugs it with silk. An adult moth emerges from an elliptical gall in late summer.

The galls are by no means safe shelters. Many of the galls you'll see in winter will have rough holes in them from downy woodpeckers or other birds pecking into the galls after the larvae. Fishermen collect ball galls to use the maggots for bait. And there are parasites that can attack gall makers inside their galls. You can never be completely sure of what you'll find when you cut open a goldenrod stem gall. You may find the gall maker, a parasite of the gall maker, an unrelated insect that took up residence in a vacated gall, or just an empty smooth-walled chamber.

Spittlebugs

Few people know what spittlebug nymphs look like despite the fact that they are easy to find. That's because the nymphs create the most unappealing shelters to hide in: big globs of frothy liquid that look like spit.

During the summer you can find spittlebugs in the tall grass of meadows, along roadsides, and in fields of alfalfa and some other farm crops. Just look for a glob of spittle perhaps an inch or so long attached to a blade of grass or the stem of some other plant.

Inside the spittle is a little insect, the immature or nymph stage of the spittlebug, sucking away at the plant juices. Nutrients are digested from the plant juices. As the excess sap is excreted, it is mixed with some special fluid from a gland in the nymph's abdomen and then whipped into a froth by a couple of little appendages attached to the nymph's rear end.

The spittle definitely protects the nymphs. Without the moist shelter, they would dehydrate and die. They can, however, create more spittle fairly quickly, so if you want to take a look at a nymph, simply wipe the spittle away with your fingers. An exposed spittlebug nymph will often turn head down and start whipping up more spittle right before your eyes. If you don't want to touch the spittle, you can poke it with a twig, and the nymph will often show itself and perhaps even jump to the ground. But if you don't want to touch the spittle with your fingers, then you should realize what good protection the spittle really is—it is even helping protect the nymph from you.

The froth doesn't provide complete protection from predators. Wasps, for example, will venture into the spittle after the nymph. The only defense nymphs still have then is to jump to the ground out the other side of the spittle. After landing on the ground, spittlebug nymphs crawl to another plant and climb up until they find a suitable spot to chew into the stem and start a new mass of spittle.

Only the nymphs live in spittle, and they are easy to identify. The adult spittlebugs are rather ordinary little brownish bugs hopping around and eating leaves. The adult's froglike appearance and habit of hopping has given rise to the spittlebug's other name, froghopper—a descriptive name, but one that hardly helps in the difficult task of distinguishing adult spittlebugs from the many other small hoppy bugs, such as leafhoppers, plant hoppers, and treehoppers.

Dragonflies and Damselflies

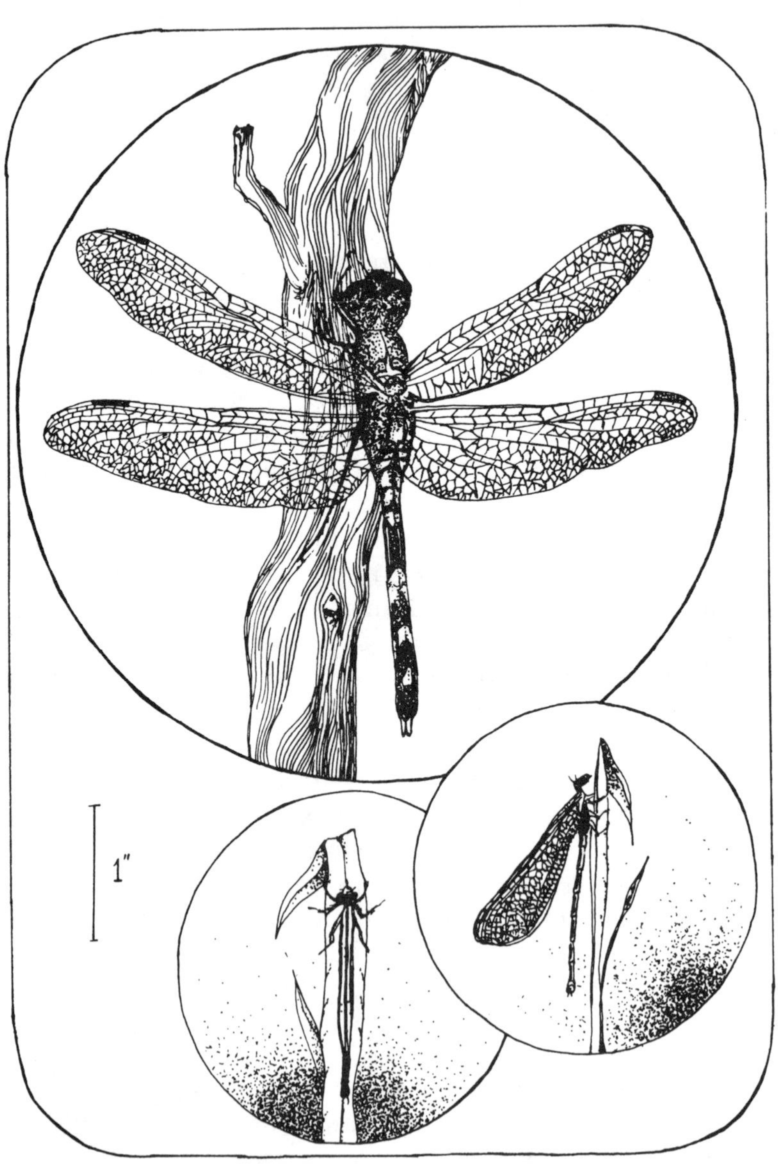

You can see dragonflies and damselflies flying over fields and meadows, but the best place to find them is near the shore of a quiet body of water. On a warm summer afternoon the shore of a small lake or river will seem very peaceful, yet all around you smaller creatures are locked in life-and-death battles. For fast-paced action, the aerial battle to watch is the one waged by dragonflies and their cousins the damselflies againt other small insects. Dragonflies will fly out on patrol like miniature search-and-destroy helicopters. They'll hover. They'll fly back and forth searching for prey. And they'll make high-speed escape flights that can hit 60 mph.

Dragonflies and damselflies have a lot of characteristics in common. Neither bite people or have stingers. They come in a variety of colors, such as green, brown, or blue. They both hunt flying insects such as mosquitoes, gnats, and even bees, which they catch by grabbing the smaller insects in the air with all six legs, forming a cage around their prey. Both dragonflies and damselflies have large, sharp eyes and four stiff wings that look like thin membranes stretched over netlike frames and that shimmer in the sunlight. And they both lay their eggs in or near water, because after the eggs hatch, the first stage of their lives is as nymphs swimming underwater, eating mosquito larvae and other small creatures. You can sometimes see dragonflies flying low over water and dipping their abdomens in the water as they lay their eggs.

Many of the differences between dragonflies and damselflies are a matter of degree. Damselflies are slower-moving, seem to tire more quickly and therefore land more frequently, and are generally smaller and daintier than dragonflies. There is one foolproof way to tell them apart: When at rest, damselflies fold their wings together over their backs the way butterflies frequently do, while dragonflies keep their wings spread out.

Dragonflies are very alert and usually won't let you get very close to them. Damselflies, on the other hand, will frequently fly right up to a person and land nearby, perhaps on a fishing pole. If you get a chance to look at one of them close up, you should examine the wing through a magnifying glass. The wings may seem clear as they fly by, but under close inspection you can see each little section reflecting a rainbow of colors.

Mayflies

Adult mayflies are delicate insects with lacy wings and long tails. They only have a few hours to a few days in which to fly around, mate, and lay their eggs before they die. Adult mayflies don't even eat during their short lives. Mayflies lay their eggs in water and mayfly nymphs live in water, so look for mayflies near ponds, rivers, lakes, and streams. You may be able to find a few mayflies anytime during the spring or summer, but the massive swarms of mayflies usually occur for a short period in April, May, June, or July, depending on the locality. Mayflies generally rest quietly during the day, taking to wing as evening approaches. A bright light such as a streetlight near a pond or river may attract hundreds of mayflies.

Once you know mayflies, identification takes only a glance. Mayflies have long bodies with two or three very long, hairlike tails. They have tiny antennae. Their shimmering, membranous front wings are much larger than their hind wings. Mayflies fold their wings together and hold them erect when they rest. There are hundreds of species of mayfly in North America, and their sizes vary. The most common mayflies have a body length of about an inch (not including the tails), but some mayflies are only ¼ inch long.

Before reaching their short-lived adulthood, most species of mayfly spend a period of at least a year as nymphs crawling around on the bottom of a pond, eating small aquatic plants. At the end of the year the nymphs crawl out of the water onto plants or rocks, shed their skins, and become winged mayflies, but not yet mature adult mayflies. Mayflies are unique among insects in that they molt again after they start flying. The second molt usually takes place within a day of the first, metamorphic molt. Look around on rocks, trees, and buildings near rivers or lakes during mayfly season and you'll see the subadult skins mayflies have shed hanging on like ghostly mayflies.

The swarms of mayflies you see around lights and over fields and ponds are almost all males. A female will fly into a swarm, mate, and fly off to lay her eggs in water. Landing on a lake or even flying in close to the water is dangerous for mayflies. Many fish feed so extensively on mayflies that fly-fishermen frequently use artificial flies designed to look like them.

Water Striders and Whirligig Beetles

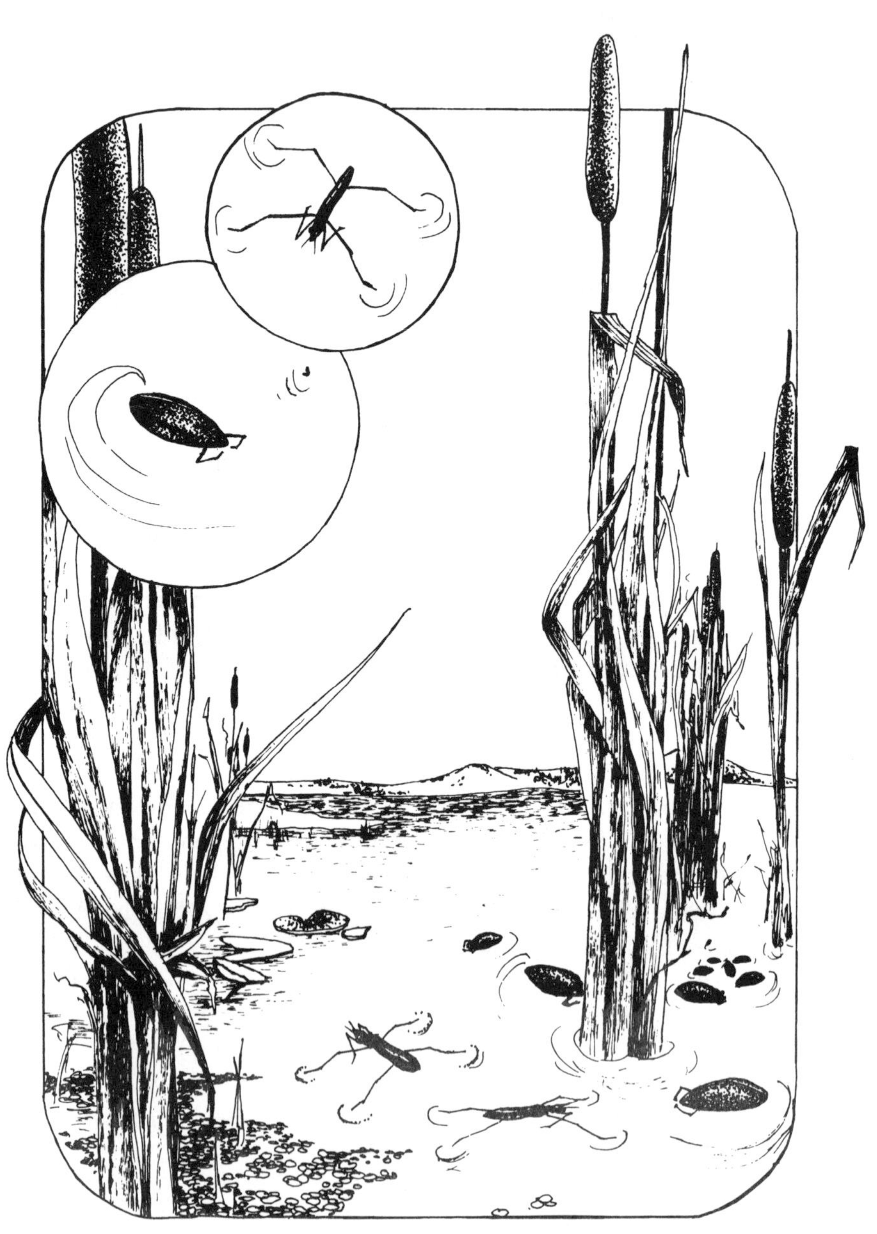

Many different insects live on the surface of lakes, ponds, streams, and marshes. They can usually be seen near the shore in the calmest part of the water, especially on warm days. Each species has a special adaptation to life on the water surface, which can be interesting to watch. Two of the most common water insects have characteristics that make them particularly easy to identify—water striders and whirligig beetles.

Water striders got their name because they actually walk on top of the water. They have four long legs tipped with short hairs that allow them to stand on the surface tension of water. The water dips down where each foot is standing, but usually not enough to break the surface tension. (Occasionally they do break through the surface tension. When this happens, they must swim to a plant or the shore, climb out, and dry off before venturing out on the water again.) From this vantage point a water strider can detect gnats and other small insects caught in the surface tension, by the ripples the gnats send out. The water strider then skates over, grabs the gnat with short front legs, stuns it, and sucks its juices out, in much the same manner as spiders consume their prey. Most water striders are about ½ inch long.

Whirligig beetles are dark, shiny, ¼ inch to ½ inch long, almond-shaped beetles that, as their name implies, whirl and gyrate around on the water surface. They are alert and active and won't let you get very close. When frightened, they will whirl around very fast in small curves. Sometimes they also dive and swim underwater, or even fly.

Whirligigs have excellent eyesight. Their eyes are especially constructed so that they can look both above and below water. They also have an echolocation system that senses wavelets bouncing back from the waves they send out when they whirl, so spinning around when frightened is actually a way for them to "look" around carefully. Whirligigs are scavengers, and wave-echolocation must be helpful to them when they feed at night. During the day they tend to congregate. You may see large numbers of whirligig beetles congregating during the spring, late summer, and fall throughout the country. Whirligig populations hit a low point in early summer and midsummer, before that year's young have changed from aquatic larvae into beetles.

Centipedes and Millipedes

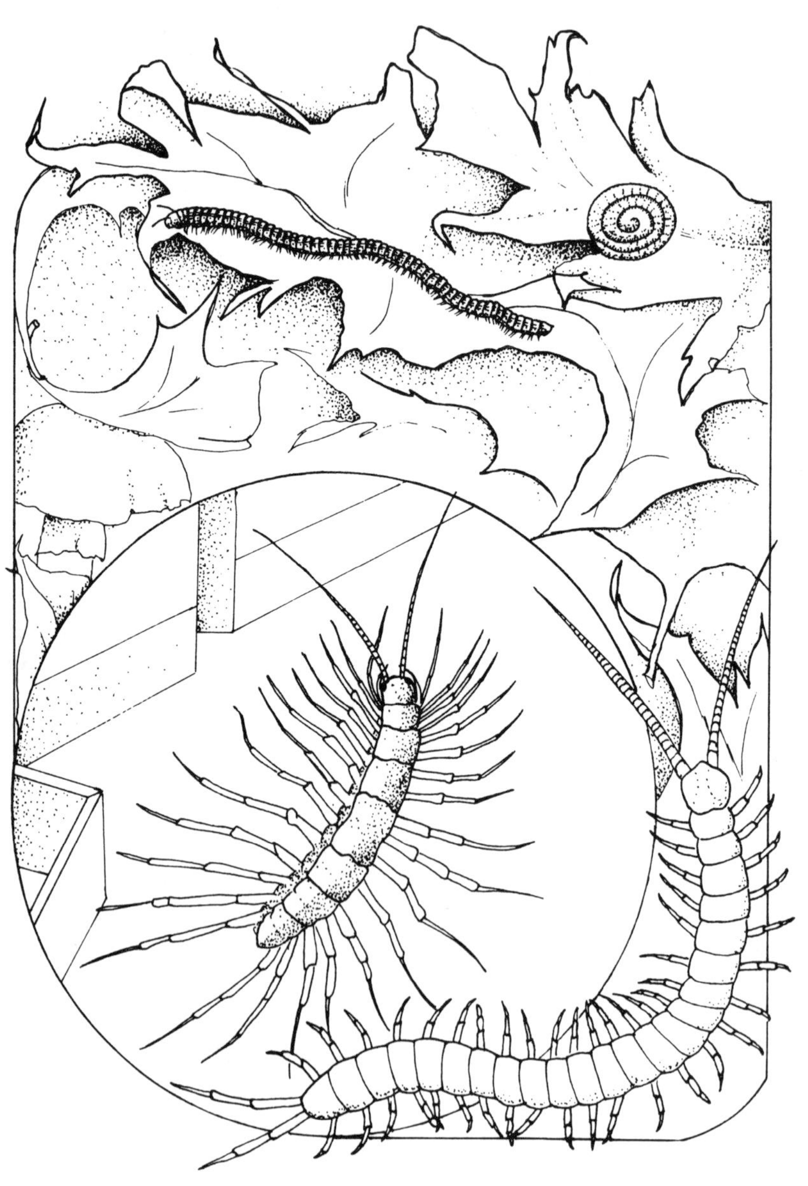

Centipedes and millipedes are wormlike creatures with segmented bodies and lots and lots of legs. In Latin, *centipede* means hundred-footed and *millipede* means thousand-footed. These names are slight exaggerations, since actual foot counts vary from about 30 to 230. Fortunately, you don't need to count all the feet to identify these creatures. You can identify them by counting the legs per segment: Centipedes have one pair of legs per segment, while millipedes have two pairs. But you'll find it easier to identify them just by watching them run: Centipedes run quite fast, while millipedes are slow.

You'll usually find centipedes and millipedes walking along the ground across paths and sidewalks. They are especially common in forests, since the shade of the trees and dead leaves keep the soil moist, and centipedes and millipedes need a moist environment to keep from drying out. You can also look for centipedes and millipedes hiding under leaves, bark, and rocks. During dry spells they burrow into the ground to keep from drying out. Centipedes and millipedes live in many other places besides forests, including in or near buildings in a city. You'll see centipedes and millipedes out and about during the day, but they are mainly nocturnal.

Centipedes are carnivorous and run to catch their prey, so they have to be fast. The first pair of centipede legs are poison claws that can hold on to an insect and inject venom. Like wasp venom, North American centipede venom is strong enough to kill a worm or an insect, but isn't normally dangerous to humans. Some of the giant centipedes that live in the tropics, however, can be life-threatening to humans.

Millipedes' diet consists mainly of decaying vegetable matter. Sometimes they'll eat delicate parts of live plants, but most of a living plant is too tough for a millipede's weak mouth. Millipedes can't run very fast, but their legs are good for pushing and burrowing into leaf mold on a forest floor.

The millipede's cylindrical body is protected by a light armor coating and stink glands. The classic defense posture of a millipede is to curl itself up in a circle or a spiral, secreting chemicals from its stink glands as it curls. This may be helpful, but predators, especially toads and starlings, still eat a lot of millipedes.

Daddy Longlegs

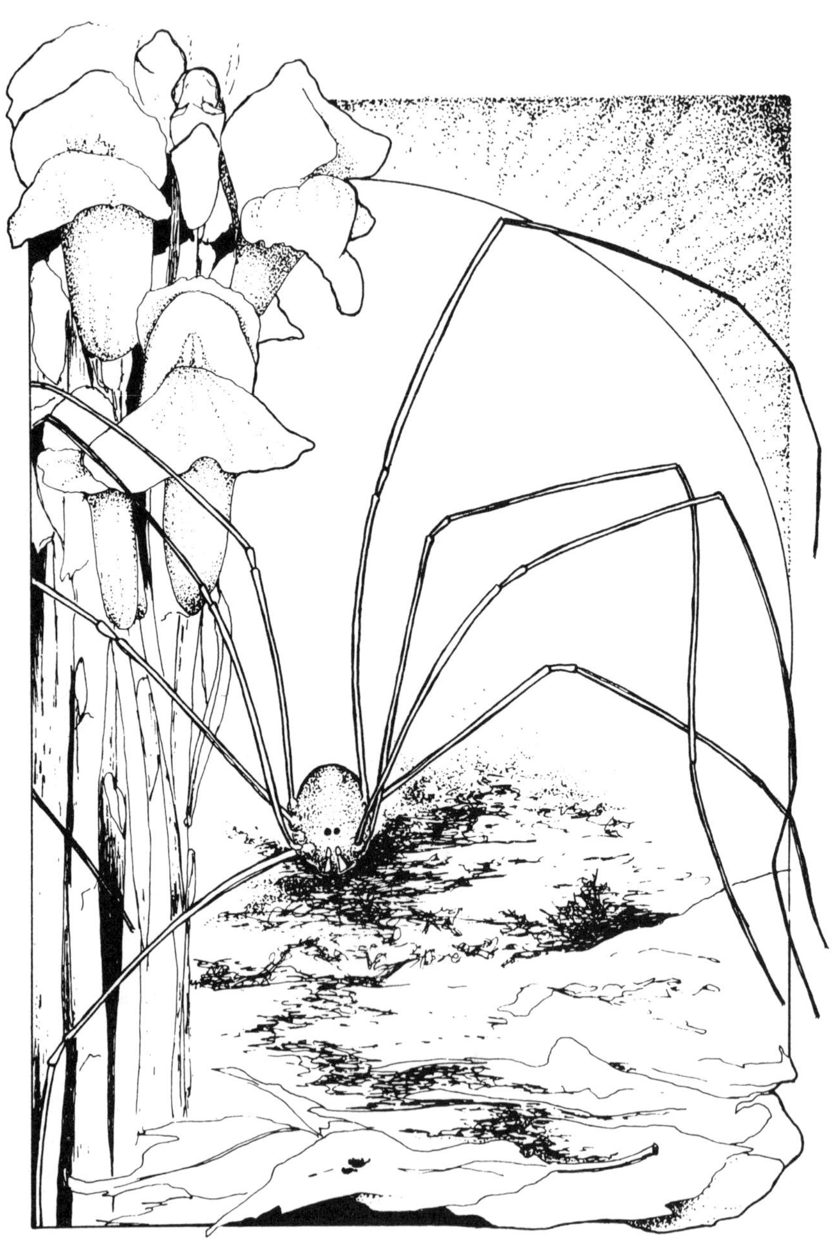

Daddy longlegs aren't insects. You can tell that because daddy longlegs have eight legs and all insects have six legs. Spiders have eight legs, but daddy longlegs aren't spiders, either. All true spiders have a "waist" between their thorax and abdomen, and daddy longlegs don't, though they are related to spiders. A daddy longlegs' body is a compact oval usually about the size of a pea, but may be considerably larger or smaller depending on the species. You identify daddy longlegs by their eight very long, very thin legs. A daddy longlegs holds its body fairly close to the ground, with its long legs arching up higher than the body.

According to an old wives' tale, if you step on a daddy longlegs it will rain. There is a tiny core of truth behind this saying. Daddy longlegs are mainly nocturnal, usually venturing out when the sky is dark, so you are most likely to see one on dark, overcast days when rain is likely. You see daddy longlegs on sunny days too, just not as often. The best time to look for daddy longlegs is in autumn, though you can find them all summer long. In England daddy longlegs are known as harvestmen, because they are most common at fall harvesttime.

Daddy longlegs generally live in woods, tall grass, and other dense vegetation. You are more likely, however, to see a daddy longlegs on open ground, such as a path or sidewalk near such vegetation or climbing a shady wall of a building, than to spot one in the dense vegetation itself. Daddy longlegs are attracted to decaying organic matter, so you can also find them around garden compost piles, perhaps hiding under nearby flowerpots or plant litter.

Daddy longlegs wander around in search of food. They don't spin webs and usually kill only small insects such as aphids, leafhoppers, and flies. They also feed on dead animals and occasionally plants.

Some of the daddy longlegs you'll see may have a leg or two missing, attesting to the dangers they face. Their major enemies are centipedes and small birds. About the only protection they have, besides running and hiding, is a pair of stink glands located near the base of the first pair of legs. Their second pair of legs have sense organs on them, and you may see a daddy longlegs wave these two legs in the air when it is disturbed.

Black Widow Spiders

You'd think a poisonous spider with such a notorious reputation would be gigantic. But black widows aren't. The body of a black widow spider is only ½ inch long, about the size of a large tack.

Black widows can be recognized by their shiny, coal-black color; their round, bulbous abdomen; and their long, wiry legs. Black widows are famous for having a little red hourglass on the underside of their abdomen, but don't rely on that for black widow identification. Sometimes the hourglass looks more like a bunch of dots, and its color may vary from red to yellow. But the biggest problem with using the red hourglass for identification is that the hourglass is on the underside of the abdomen and therefore often hidden from view. With black widow venom so dangerous, you don't want to poke around trying to get a better view. It is much better just to assume that any ½-inch-long or smaller black spider with a big abdomen is a black widow.

Black widows are seldom found out in the open. They weave their irregular webs in sheltered places such as along the ground among fallen branches. Now that humans are building all sorts of nice sheltered places, the black widow has shown proper appreciation and moved into dark corners of basements, garages, and attics. Just about any type of shelter can attract a black widow, from a park bench to a flagstone wall.

Black widow spiders are most common in the southern United States, especially in late summer and fall, but they are not limited to the South. Black widows or their close relatives such as northern widows can be found from Florida to California and as far north as southern Canada.

Male black widow spiders are considerably smaller than females and lack some of the easily identifiable characteristics, such as the bulbous abdomen. But since males aren't dangerous, black widow identification is concentrated on females. And the females do bite. They are not aggressive attackers, but they will defend their webs, particularly when egg sacs are in the web.

The bite itself doesn't hurt much, and sometimes black widows inject little or no poison. But when they do inject a lot of venom, watch out. People have died from black widow bites.

Black-and-yellow Garden Spiders

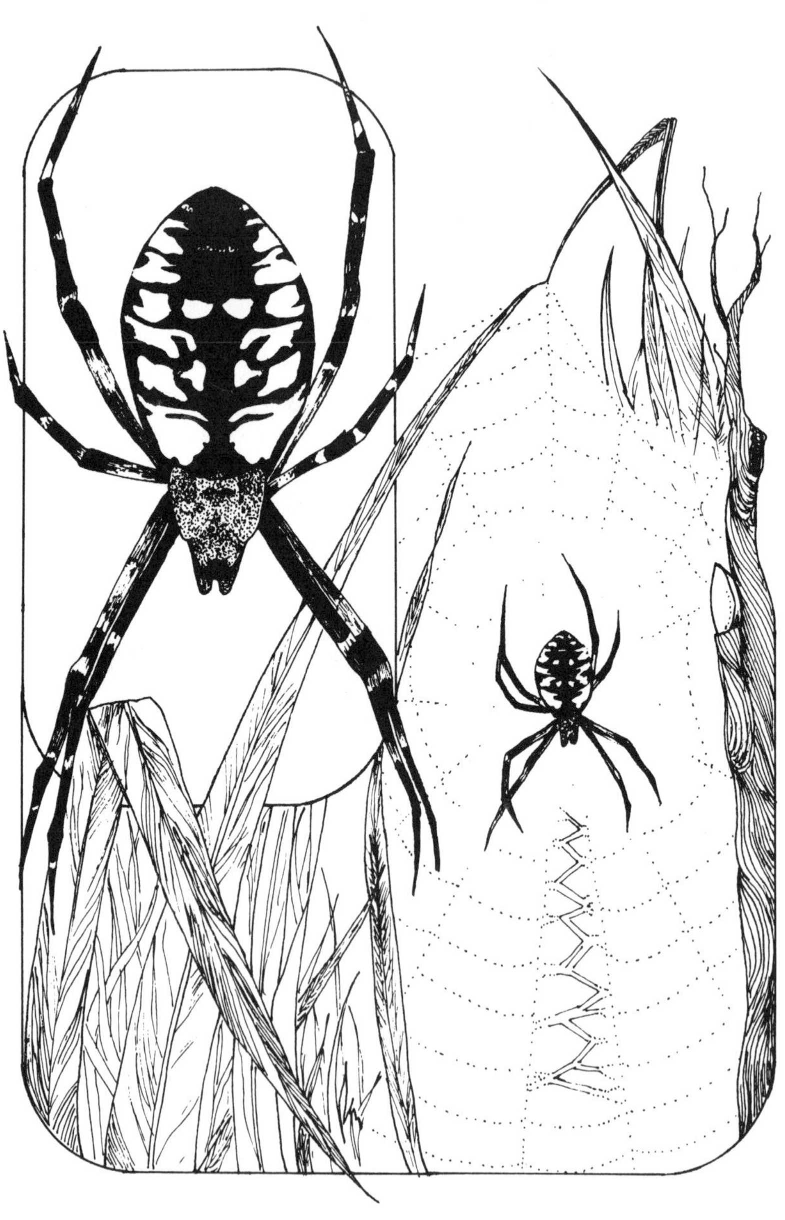

Black-and-yellow garden spiders are easy to see because their bodies are one inch long and strikingly colored jet black and bright yellow. Look for black-and-yellow garden spiders during the summer in sunny meadows that are somewhat sheltered from the wind. These spiders weave large, vertical webs between a couple of tall weeds, often right across a meadow path. You'll find the large black-and-yellow garden spiders sitting in the center of the web.

Black-and-yellow garden spiders' colors come in oddly shaped blotches. The black is always jet black. The yellow is usually bright yellow, but may be pale yellow or even orange. So long as the color is in blotches, the spider is a black-and-yellow garden spider. The banded garden spider is a similar large spider colored with black, pale yellow, and white. A banded garden spider is easily identified by the regular horizontal lines or bands covering her entire back.

Most of the black-and-yellow garden spiders you'll see will be female, because males just aren't as noticeable. The female's body is about an inch long, while the male's body is only ¼ inch long. Sometimes you'll see the male's small web near the bottom of a female's web.

To watch a garden spider in action, lightly toss a small grasshopper or other insect into the web. Black-and-yellow garden spiders don't seem to mind people watching them closely as they wrap up their prey. Their response to perceived danger is to jump to the ground and hide.

Garden spider orb webs may be as long as six feet. If you pluck a stem of grass and touch it to one of the radii of the web, you'll find that the radii aren't sticky. Touch it to the spiral and you'll find that the spiral is very sticky and stretchy. Just below the center of the web you'll notice the garden spider's signature, a zigzag patch of white silk called a stabilimentum.

Scientists have long thought that the stabilimentum helped stabilize garden spiders' webs. Recent studies suggest that its main function may be to help birds see and therefore avoid the web. Even with stabilimenta, garden spider webs are often destroyed by larger animals. How can they avoid frequent damage when they build their webs across meadow paths? A black-and-yellow garden spider usually builds a new web during the night.

Sulphur Butterflies

Sulphur butterflies are so common that many experts think the butter-yellow color of some sulphur butterflies gave rise to the name butterfly. However, many species of sulphur butterfly are not butter yellow. Sulphurs can be any shade of yellow or even orange. They will usually have a dark border to their wings, or at least some simple black markings. Some members of the same family of butterflies are primarily white instead of yellow, so instead of being called sulphurs, these butterflies are simply called whites.

Whether sulphurs or whites, all members of this family have rounded wings with a wingspan of about 1½ inches. Any medium-sized yellow or white butterfly you see that has rounded wings and some black markings will almost certainly be a sulphur or a white.

Look for them on warm, sunny summer days. They are quite common and easy to see flying above a meadow. Also, some species of sulphurs swarm around mud puddles by the dozens and allow you to approach within several feet.

The best place to find the green, hairless sulphur or white caterpillars is on cabbage or related vegetables, such as broccoli, growing in a garden. The caterpillars also live on many meadow plants, such as wild mustard, clover, and alfalfa.

While males and females of the same species have slightly different markings, you may find it more interesting to try to identify male and female sulphurs by their behavior. Obviously some of their behavior, such as eating, is the same for both sexes. But there are times when you can tell them apart. Sulphur butterflies that you see flying slowly over a field and landing briefly on leaves are probably females laying eggs. Sulphurs that you see flying rapidly up to any and all butterflies are probably male.

When a male sulphur finds a female of his own species, he will fly around and around her. If she isn't interested, the female often flies upward, with the male still circling around her. They will spiral high in the air until finally the male gives up and flies quickly to the ground. The female then descends slowly. These spiral flights of the sulphur butterfly are quite common and can be seen most of the summer. On a nice day in a meadow, you may even see a couple of spiral flights going on at the same time.

Swallowtail Butterflies

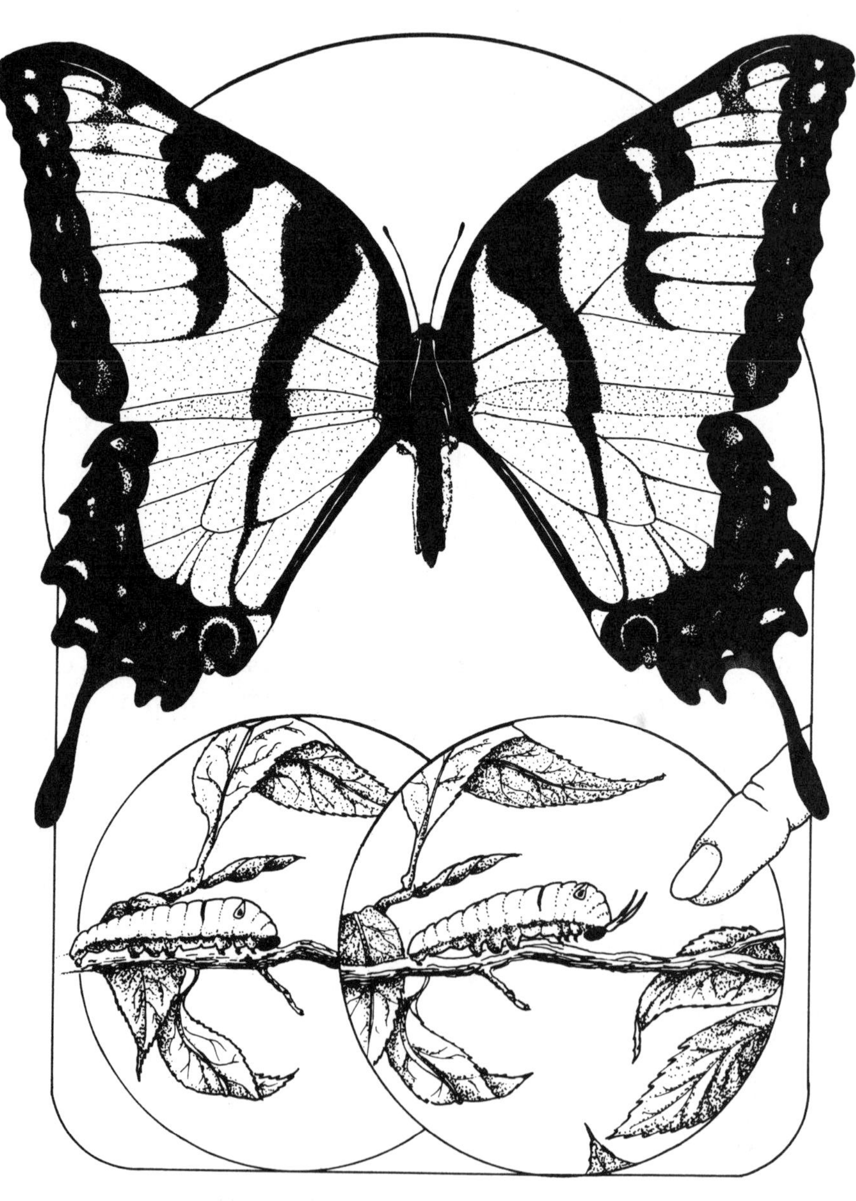

Swallowtail butterflies are quite large, with wingspans ranging between 2½ and 5 inches. Most of them are the size of monarch butterflies (approximately 3½ to 4 inches) or even larger. Swallowtails are strikingly colored with black and blue, white and black, or yellow and brown or black. They'll often have orange or blue spots as well. Whatever the color, though, you can identify swallowtail butterflies by their tails. The tails are narrow prongs sticking out on the end of each hind wing. The name swallowtail comes from the tail's similarity to a barn swallow's deeply forked tail, which ends in two prongs.

Swallowtail butterflies are attracted to bright flowers, so the best place to see swallowtails is in a flower garden. The next best place would be a sunny meadow at the edge of some woods. Not all swallowtails live in meadows; some species live in deserts and some live in woods. But most swallowtails at least visit meadows. Even woodland swallowtails visit nearby meadows. Another good place to look for swallowtails is around puddles and other sources of water. Sometimes you'll see several swallowtails sipping from the same mud puddle.

Swallowtail caterpillars are also easy to identify, because of a strange stink gland they have right behind their heads. When you touch a swallowtail caterpillar, it will suddenly flare out the forked, hornlike stink gland. The sudden flaring of this foul-smelling yellow or orange projection can startle away some would-be predators.

You'll find swallowtail caterpillars on weeds, trees, and vines. Swallowtail caterpillars are among North America's largest caterpillars, measuring as much as 2½ inches. None of them are hairy, and they come in several colors, green being most common. Many swallowtail caterpillars have two large spots on their back that look like eyes. The caterpillar's real eyes are quite small and much farther forward. The large eyespots make them look more ferocious, especially in conjunction with the stink gland's projection.

There are a couple of other butterflies besides swallowtails that have taillike projections from their hind wings, but these are either smaller or reddish. Swallowtails are never reddish.

Aphids

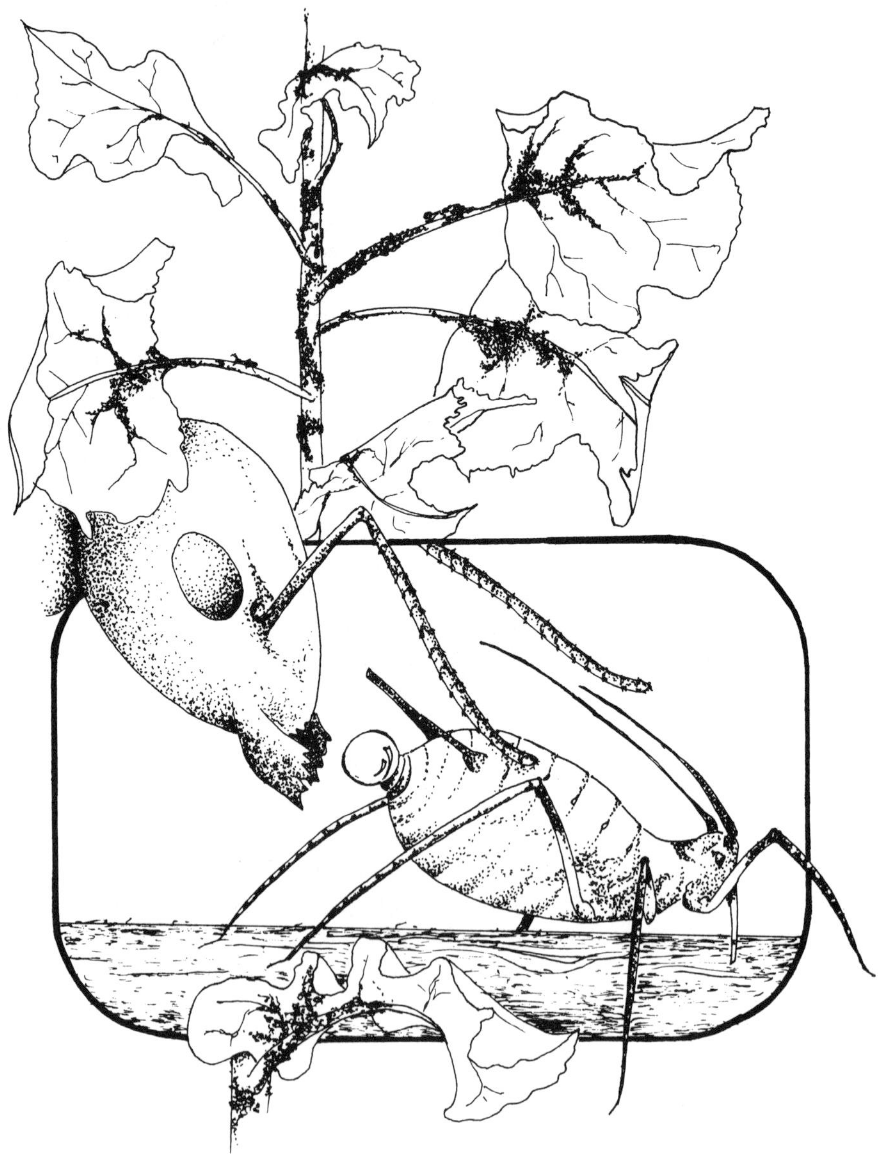

Aphids are very small insects, only ¹⁄₁₆ inch to ⅛ inch long when full-grown. Most aphids are green, but some species are brown, white, or even pink. You can easily find aphids by looking on the underside of leaves, and if you peer closely you may see their distinctive pointed snouts. During the summer aphids prefer the tender young leaves that grow at the tips of tree branches and tops of weeds. Especially check out young leaves that have curled slightly, for this is often a sign of aphid damage.

A curling leaf at the tip of an apple tree branch may have hundreds of little aphids on its underside. At first glance aphids will look like little bumps. Under close inspection, each bump turns into a tiny insect. Most aphids don't have wings. The small percentage of aphids that do have wings are able to spread to new trees and weeds as well as whole new regions if they have a favorable wind.

You'll find aphids living on stems and trunks as well as leaves so long as the stems are soft enough for aphids to poke their snouts in and suck plant sap. Plant sap, which is mostly sugar and water, is all aphids eat, so by the time aphids suck out enough sap to supply their protein requirements, they've consumed much more sugar than they need. Aphids excrete this excess sugar in tiny droplets called honeydew. Many insects, such as bees and ants, collect honeydew. Ants, however, go one step beyond simple collection. Ants will watch over and protect the aphids as well.

This ant-aphid relationship is very common. Many of the plants you'll see with aphids on them will have ants tending the aphids. Just look at the trunk or stem. If ants are tending aphids, there will be regular traffic of ants going up and down the tree trunk, almost like a highway. Ants guarding the aphids can be very protective. To test this, try moving a finger toward an ant you see among a group of aphids. Instead of running away, the ant may rear up toward your finger as if ready to attack.

Ladybugs, or ladybird beetles, and many other insects dine regularly on aphids. To survive the ravages of their enemies, aphids have an incredibly high birthrate. When only about a week old, female aphids start giving birth to live young, even without mating. Mating does take place, though, before aphids lay their eggs in the fall.

Tent Caterpillars and Fall Webworms

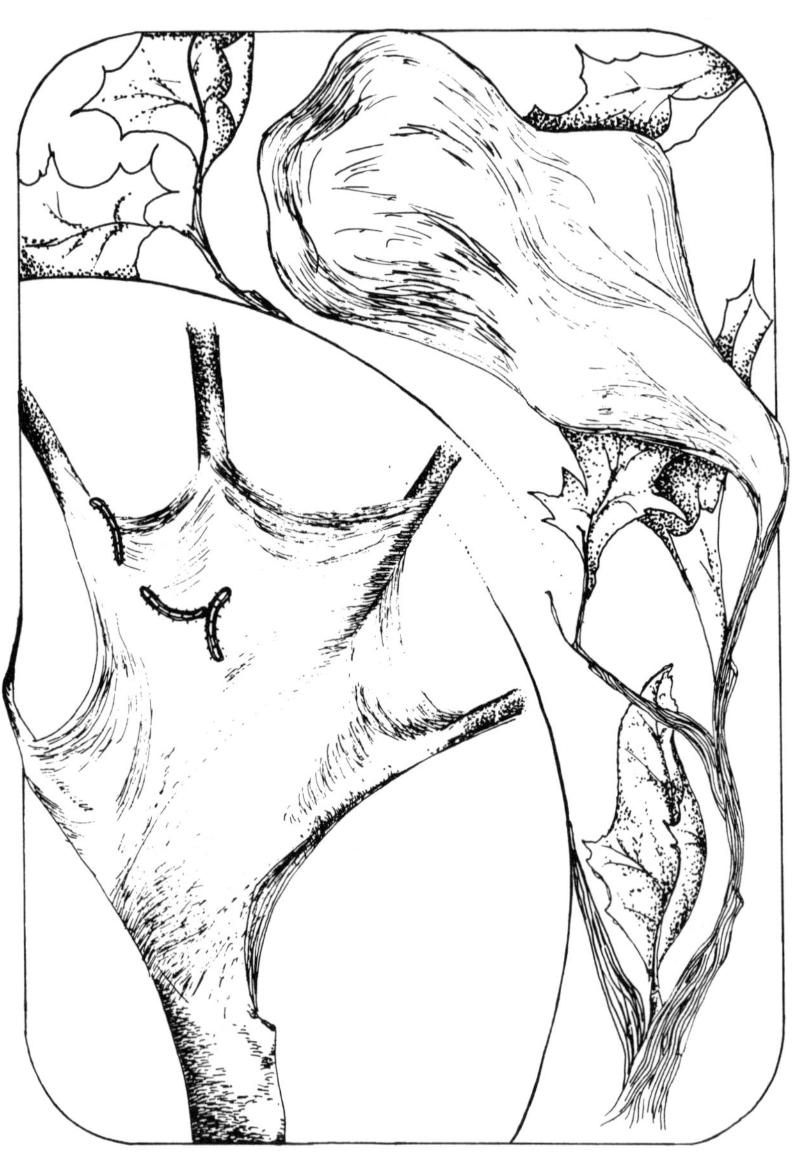

In spring, summer, or fall you may notice large, dense webs in among the branches of trees, particularly trees at the edge of woods or along roads. Caterpillars of several different moths weave such webs. The two most common are tent caterpillars and fall webworms.

The moths of both tent caterpillars and fall webworms are not very easy to find or identify. You can, however, easily identify the caterpillars simply by where their webs are located: Fall webworms prefer leafy ends of branches, and tent caterpillars prefer a tree crotch where the branches can serve as tent poles.

Look for tent caterpillar colonies in the spring and early summer. Tent caterpillar eggs hatch in the spring. The baby caterpillars stay together and weave themselves a protective shelter at the crotch of some branches. Inside this tent there is no food, so tent caterpillars must venture forth to eat their fill of leaves and then return to the shelter. If you look closely at the branches leading from the tent, you'll see strands of silk left by the caterpillars. These strands help the caterpillars find their way home and serve as trail markers to good feeding grounds.

You won't see fall webworm colonies until late summer or early fall. Fall webworm eggs are laid on leaves at the ends of branches, and the young caterpillars weave a protective web around the whole end of the branch, leaves and all. By the time fall webworms have reached their full size, their web may be a couple of feet long and enclose the ends of several branches. The web building is done at night so the caterpillars can stay in their shelter all day.

A tree can feed a few caterpillars with no difficulty, but several colonies of tent caterpillars or fall webworms can completely defoliate a small tree. The fact that both species can become numerous enough to damage trees is proof that their webs are a good defense. But the webs are not a perfect defense. Tent caterpillars must venture out to eat, and risk being eaten themselves. Fall webworms can feed within their webs, but when they get too close to the edge they are sometimes picked off by small birds or hornets. And some larger birds, such as cuckoos and orioles, will snatch tent caterpillars or fall webworms from within their protective webs.

Banded Woolly Bear
Caterpillars

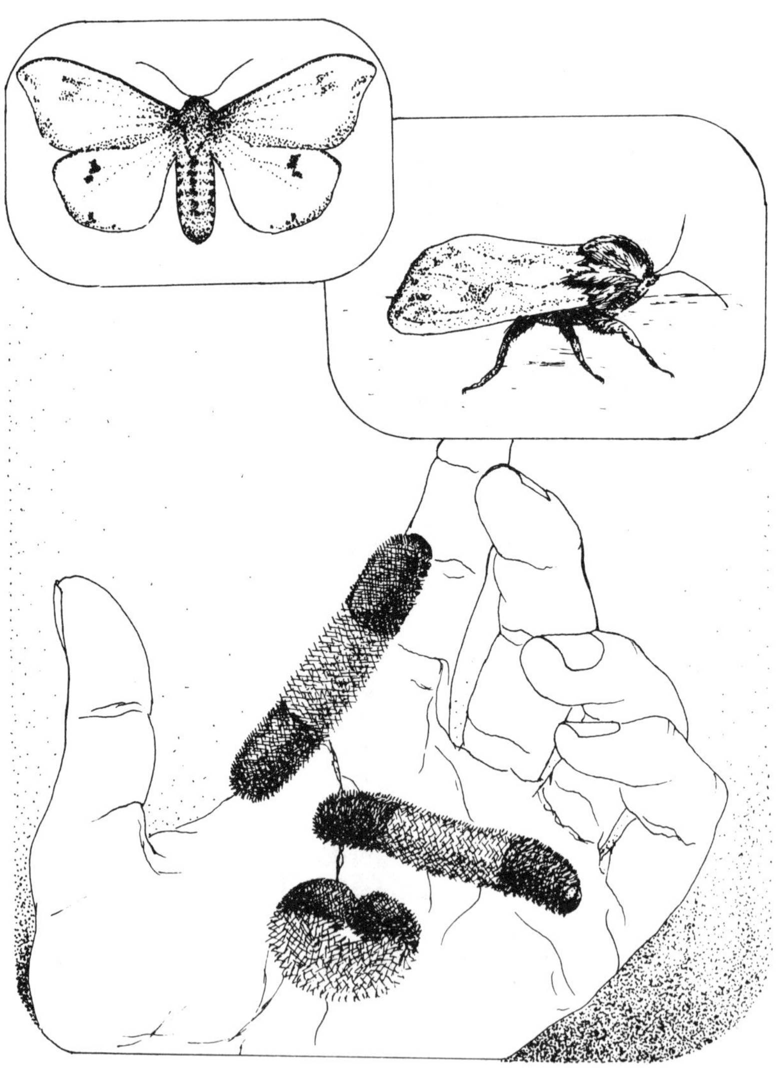

Hardly anyone pays any attention to the Isabella moth, yet the caterpillar stage of the Isabella moth is written and talked about in popular media every fall, because the moth's caterpillar is the banded woolly bear, which is said to predict winter weather.

Look for banded woolly bears crossing roads and paths on warm autumn days when the leaves are turning color and falling. You'll notice that banded woolly bears seem to spend an awful lot of time just walking. Perhaps "jogging" would be a better description, since many of them move along faster than a simple caterpillar walk. To catch a woolly bear, just place your hand on the ground in front of it. It will usually walk right up onto you. Frightened woolly bears will curl up into balls and stay that way until they think it's safe.

There are several species of large, active, furry caterpillars that go by the name woolly bear, and they produce different moths. The banded woolly bear is about 1½ to 2 inches long and almost entirely covered with dense ⅛-inch-long hair. The hair length is very uniform, but the hair color varies to give this woolly bear three distinct color bands: black in the front, brown in the middle, and black again in the rear.

Some people say that the width of the different colored bands on a banded woolly bear will predict how cold the winter will be. Normally the banded woolly bear's central brown band averages about as wide as the two black bands combined. According to this old folk tradition, the winter will be mild if the brown band is wider than normal, and cold if the brown band is narrower than normal. However, scientific studies haven't shown any connection between woolly bear coloring and winter weather.

Unlike most caterpillars, banded woolly bears must be concerned with winter weather, because they must survive the winter as caterpillars. The woolly bears you see in the fall don't weave cocoons until after they've eaten some fresh weeds in the spring. Much of the banded woolly bear's incessant walking seems to be a search for a proper place to hibernate. Woolly bears hibernate where they might be protected from swings in temperature as well as predators, such as under fallen bark. Warm weather in winter can rouse a banded woolly bear and have it out marching again.

WEATHER SIGNS

Spring Air Wars

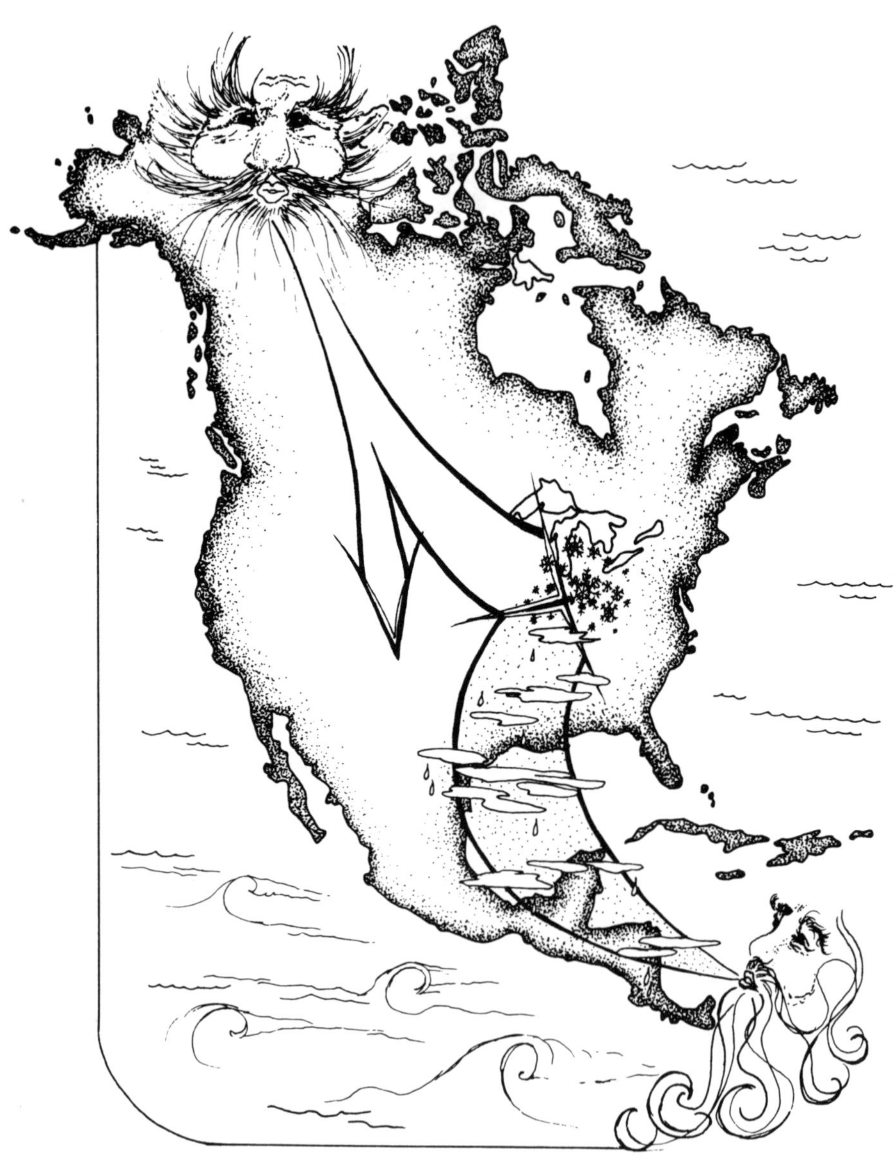

Early spring weather for most of North America east of the Rocky Mountains is primarily a war between cold, dry northern air and warm, moist southern air. You can usually tell which air mass is winning simply by looking at the sky in the morning. If the sky is clear, northern air is winning. If the sky is cloudy, southern air is winning. If it is snowing, watch out! You're in the battle zone, and spring snowstorms can be the worst storms of the year. Spring snowstorms frequently dump more snow than midwinter storms, because spring air is warmer and warmer air can hold more moisture, offering the potential for a major snowstorm.

The northern air that swoops down on us is part of the continental polar air mass. This air is known for being both cold and dry. As it moves south, the northern air warms somewhat, which allows it to hold more moisture without producing any clouds. Thus, if the sky is clear in winter or early spring, you are probably under the influence of the continental polar air mass.

The southern air that sweeps up on us is part of the maritime tropical air mass. Formed over warm water, this air mass holds a lot of moisture. As the air moves north over land, it usually cools enough to form a light cloud cover, particularly at night and early morning. Sunlight may evaporate the clouds by afternoon, so to tell which air mass is winning the spring air wars you should check the sky early in the morning. If it is cloudy, you are probably under the influence of the maritime tropical air mass.

A Pennsylvanian folk tradition has it that you can predict the rest of the winter according to which air mass is winning the spring air wars on February 2 (though most meteorologists disagree). We'll have an early spring if a groundhog, also known as a woodchuck, comes out of its hole the morning of February 2 and doesn't see its shadow, due to maritime tropical clouds. We'll have six more weeks of winter if the groundhog does see its shadow, thanks to clear continental polar air.

Interesting problems arise when people try to apply a local weather tradition nationwide. For instance, six more weeks of winter after February 2 may mean a late spring in Pennsylvania but an early spring in Minnesota.

Clouds

The ten main types of cloud are identified by five Latin words in various combinations: cumulus, stratus, cirrus, nimbus, and alto (from the Latin *altus,* or high).

Cumulus means to heap or pile up, as in the word *accumulate.* Cumulus clouds appear as fluffy piles. *Stratus* means to stretch out, and refers to layerlike or sheetlike clouds that stretch out. *Cirrus* means curl or filament and refers to very high, thin, wispy clouds. *Nimbus* means rain cloud. A stratus cloud may produce a little light drizzle or snow flurries, but if the cloud is dark enough to really rain or snow you add *nimbus* to its name. *Alto* modifying a word means high but not too high. For example, an alto voice is a high voice for a man but low for a woman.

You combine these five words to name the ten major types of clouds. There are three very high, thin clouds: cirrus, cirrocumulus, and cirrostratus (see page 104 for more on cirrostratus clouds); two middle clouds: altostratus and altocumulus; and three low clouds: stratus, nimbostratus, and stratocumulus. Stratocumulus is a strange combination of terms referring to sheet clouds that aren't uniformly gray. They have light and dark areas, indicating some heaping. Cumulus clouds are the puffy, cotton-ball clouds. When they pile up quite tall, they are called cumulonimbus. Some cumulonimbus clouds grow extremely tall and their tops spread out flat like anvils. Cumulonimbus clouds can produce rain, snow, hail, and lightning and thunder.

When lightning sizzles through the sky, the air it passes through is superheated and expands as rapidly as if there had been an explosion. The sound waves produced by the suddenly expanding air result in thunder. You can tell how far away a lightning bolt is by watching a flash of lightning and then counting the seconds until you hear the thunder. The speed of light is so fast that for practical purposes you see the lightning the moment it flashes. The speed of sound in air is $\frac{1}{5}$ mile per second. So if you see a flash of lightning and count five seconds until you hear the thunder, the lightning bolt was a mile away.

Highs and Lows

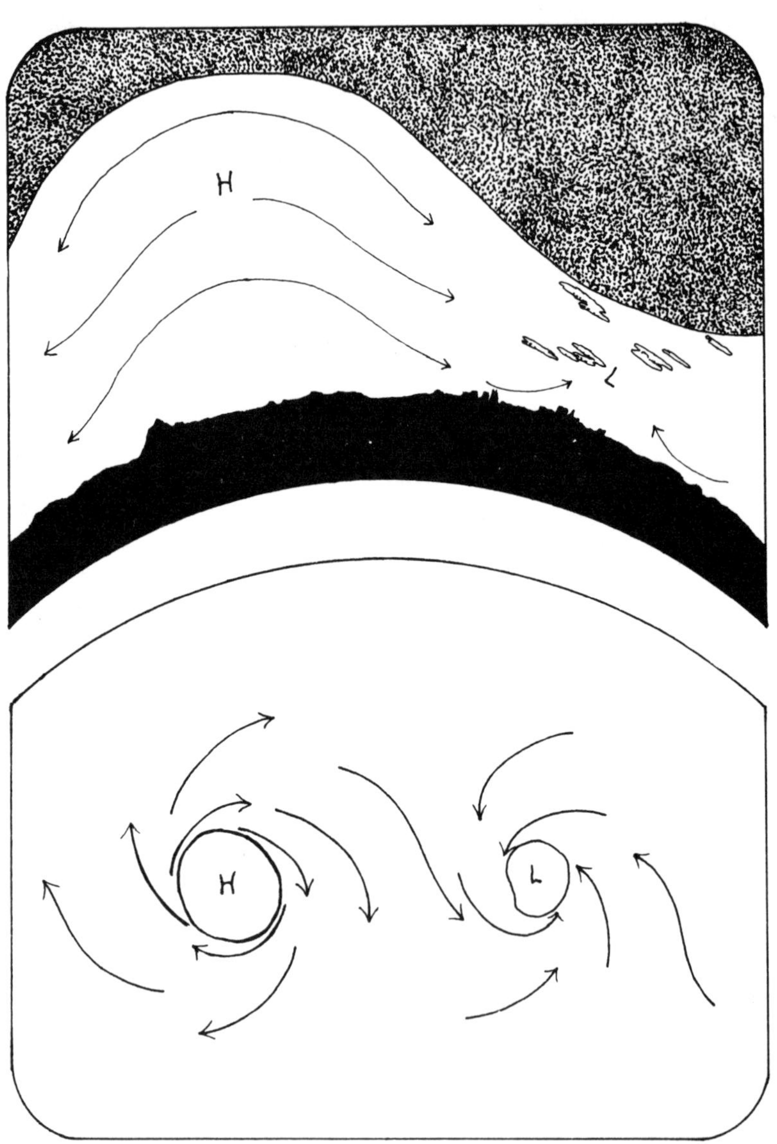

You don't need weather maps or barometers to identify high- and low-pressure systems. All you need to do is study the weather. Rain or snow positively identifies the presence of low barometric pressure. Generally, cloudy and windy conditions are also indications of low pressure. Highs are recognized by the presence of fair weather, light winds, and few or no clouds.

Atmospheric pressure is simply the weight of the air. The typical high-pressure system is a thousand-mile-diameter mass of heavy air that spins slowly clockwise. Low-pressure systems come in many forms. Some lows are circular masses of light, rising air three hundred miles in diameter and rotating counterclockwise. Lows associated with thunderstorms may be twenty miles in diameter. At the leading edge of warm and cold fronts you can find very long, narrow troughs of low pressure. To complicate identification further, some lows that form over deserts may not even have enough moisture to form clouds.

To understand what is going on inside highs and lows, think of yourself as living on the bottom of a vast ocean of air. Huge waves and troughs pass over the top of this ocean of air.

Underneath a wave, the barometric pressure would be high. Air will tend to move down and away from this high-pressure wave. As the air comes down, it warms and can therefore hold more moisture: hence few or no clouds and no rain.

Underneath a trough, barometers would show a low. Air from surrounding regions moves into the low as the air in the center of the low rises to fill in the trough. Air cools as it rises, and this forces any water vapor to condense into clouds and rain or snow.

The spin of highs and lows is caused by the rotation of the Earth under moving air, in much the same way that the prevailing westerlies are created (see page 100). Highs and lows rotate in different directions because air is moving out of highs and into lows. That is also why highs have mild winds and lows have stronger winds. Air spiraling out of a high gets to spread out and slow down, whereas air spiraling into a low is constricted (by more air spiraling into the low) into smaller and faster circles.

Prevailing Westerlies

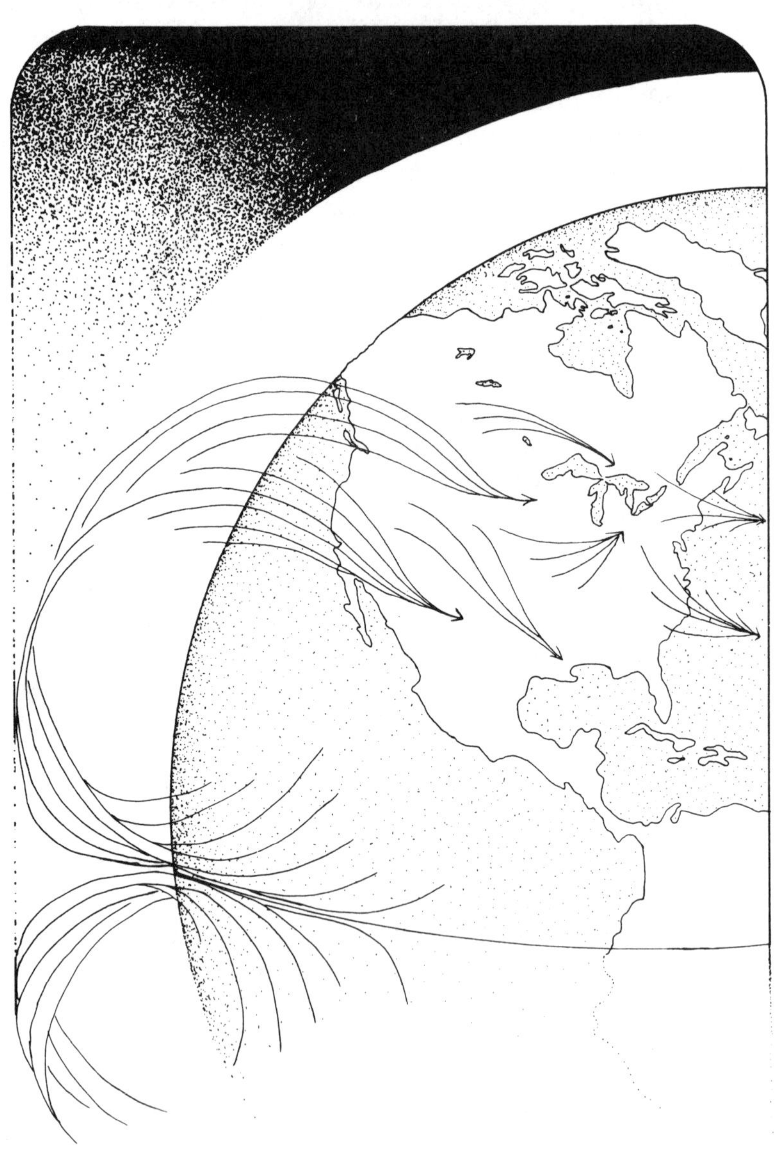

Which way does the wind blow most of the time? We are part of a dynamic and vital planet, and it is fascinating to be able to identify its most familiar motions. If you live anywhere north of New Orleans on the Gulf of Mexico and south of Churchill on Hudson Bay, you are living in the belt of the prevailing westerlies, an area that stretches all the way around the globe from about 30 degrees latitude to 60 degrees latitude, where west winds are said to prevail. Winds won't always be from the west, but if you pay attention to your weather you'll often be able to recognize the influence of the prevailing westerlies.

The easiest kinds of prevailing westerlies to identify are surface winds blowing from the west, northwest, or southwest. Also watch which way storms move when you can see them starting. Did the prevailing westerlies blow the last storm you had off toward the east? The paths storms take aren't always easy to see, but are almost always affected by the prevailing westerlies. Even storms that blow in off the Atlantic are eventually deflected by the west winds. Blizzards from the north and rainstorms from the south are often deflected until they seem to be coming from the northwest or the southwest.

The west winds that blow into, say, Chicago actually get their start in the calm air near the equator. As the intense tropical sunlight warms the air, the air rises. When this warm air reaches the top of the troposphere, it moves north and south. This air cools as it gets closer to the poles, and part of it descends into the lower atmosphere, creating the prevailing westerlies.

How can calm air move north and create strong west winds? At the equator, the calm air is moving at the same speed as the Earth. As the Earth spins, a spot on the equator sweeps a much bigger circle than Chicago does—so much bigger, in fact, that a spot on the equator is moving 250 mph faster than Chicago is. The once-calm air at the equator retains some of this faster west-to-east speed as it moves north. As this air cools and mixes with local air, it creates west winds.

In the middle of the North American continent, west winds are by far the most common winds. In other places, particularly along a shore where sea breezes are common, you may not see surface winds from the west as often. But if you are in the belt of the prevailing westerlies and keep your eyes open, you'll frequently be able to see the effects of the prevailing west winds.

Warm Fronts and Cold Fronts

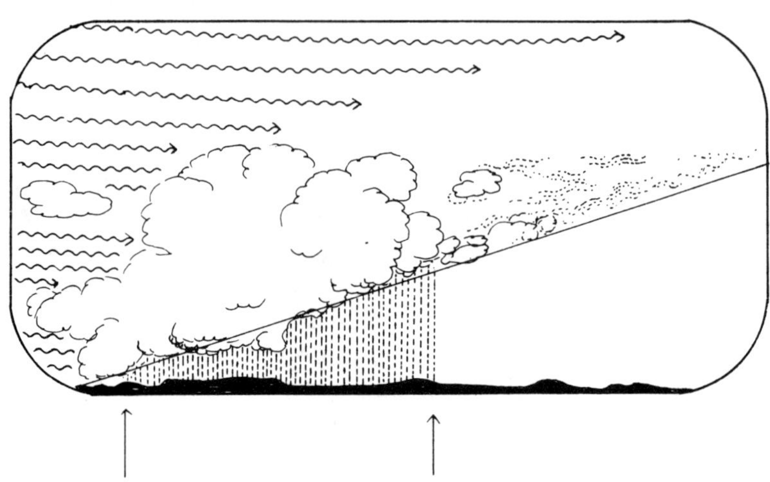

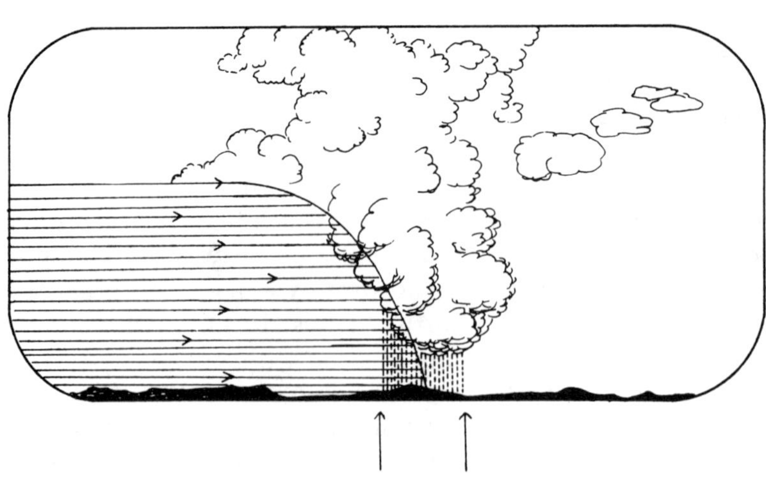

The next time a meteorologist tells you that a warm front or a cold front is coming your way, pay attention to your weather. Fronts are the leading edges of large air masses that may be thousands of miles in diameter. These air masses are continually sweeping across North America, so the storms associated with fronts are important aspects of our weather. They don't always produce storms, but you may be surprised how often you can easily tell warm fronts from cold fronts just by the way it rains. A typical warm front produces mild, sometimes intermittent, rain, snow, or drizzle for twenty or thirty hours. A typical cold front produces strong rain or snowstorms that last less than half a day, often only three or four hours.

Cold air masses are obviously generated up north and warm air masses down south. But the factor that determines whether a particular mass of air behaves as a cold or a warm front is whether the air mass is colder or warmer than the air it is displacing. And the air mass doesn't have to be very many degrees colder or warmer to create typical frontal weather.

Warm air rises above cold air, so a warm front pushing its way across North America has a tendency to slide above the air it is displacing. The leading edge of a warm front, then, is high above and can be hundreds and sometimes even thousands of miles ahead of the line where the warm front meets the ground. As air rises, it cools, and colder air can hold less moisture. So as the warm front air slowly moves up to the high leading edge, it releases moisture. Because the air is rising so slowly, its moisture is released only gradually, spreading gentle rainfall over a wide area.

A cold front has a very thick leading edge that burrows under the warmer air it is displacing. Much of the air in the path of a cold front is pushed off to the sides of the air mass. Some of the wamer air, however, is forced up over the leading edge of the cold front. As this warmer air is forced to rise rapidly above the thick leading edge of the cold front, it cools and releases its moisture. The rapid rise means that most of the air's moisture is released quickly, creating a narrow storm that is raining or snowing fairly hard.

Imagine a combination of a storm with a narrow range of precipitation traveling, typically, faster than a warm front, and you'll see why cold-front storms don't normally last as long as warm-front storms.

Lunar Halos

You're outside on a moonlit night, you glance up, and there is the moon shining brightly, but around it is a ring of light, a glowing ring that is some distance from the moon itself. Old folk traditions say this means rain is on the way. And some modern meteorologists agree, giving this rule of thumb: On two out of three occasions, it will rain or snow within twenty-four hours following a lunar halo.

The halo is caused by very high, very thin cirrostratus clouds. These clouds are so thin that if it weren't for the halo, you might not even notice them. They are about six or seven miles up and composed entirely of ice crystals. As moonlight passes through the cirrostratus clouds, the ice crystals act like prisms and refract the light. When you look at a spot on the halo, you are seeing moonlight that might have landed far away if the ice crystals hadn't bent it over toward you.

The reason cirrostratus clouds imply precipitation is that most cirrostratus clouds are formed by warm fronts. And warm fronts (see page 102) are usually associated with rain or snowstorms. The high cirrostratus clouds are at the leading edge of warm fronts, far ahead of the storms. Usually the rain or snow won't get to you for at least six hours after you see the lunar halo. Occasionally the precipitation will take as long as forty-eight hours to reach your area. Remember, though, that there are exceptions to all general weather rules.

You'll see lunar halos sooner or later if you frequently look at the night sky. If you particularly want to see one, pay attention to weather reports. When the meteorologist says a warm front is heading your way, make a special point of looking for lunar halos. A full moon is best for this, because it rises at sunset and is visible all night. But even a crescent moon can create a bright lunar halo. Cirrostratus clouds also form halos or a similar effect around the sun, but because they occur in bright daylight, solar halos are often not as obvious or as strikingly beautiful as lunar halos.

Moonlight passing through lower clouds that are made up of water droplets may create a fuzzy light area around the moon. This type of light area is called a corona and will always appear to be touching the moon. You won't confuse coronas with halos if you remember that halos never appear to be touching the moon.

NIGHTTIME PHENOMENA

The Big Dipper and the Little Dipper

The Big Dipper is the easiest constellation to recognize. It is made up of bright stars that actually form the outline of a standard, Colonial-style water dipper. And you can see it all year round.

To find the Big Dipper, look in the northern part of the night sky and remember that it is big. The Big Dipper is close to the North Star, around which all stars appear to rotate. Sometimes you'll see the Big Dipper high in the sky but upside down. Other times it will be right side up close to the northern horizon. And you'll see it tipped in all the angles in between, perhaps even losing part of its handle over the northern horizon.

The two stars in the Big Dipper farthest from the handle are called the "Pointer Stars," because a line drawn through them and extended up (from the Dipper's point of view) six times the distance between the Pointer Stars will bring you to the North Star, Polaris. Actually, the Pointer Stars don't line up directly with Polaris, but the line comes so much closer to Polaris than to any other star that making a mistake is difficult.

Polaris is not extremely bright. Its fame comes from the fact that it always points due north, more accurately than most compasses. Polaris is also the end of the handle of the Little Dipper. Even knowing where the Little Dipper starts, you can still have a hard time seeing this constellation because several of the stars in the Little Dipper are so faint that they disappear in moonlight or city lights, and the "dipper" that the Little Dipper forms is badly bent.

A pre–Civil War song urged runaway slaves to "follow the Drinking Gourd" north to freedom. You, too, may find the Little Dipper easier to see if you look for a crookneck drinking gourd rather than a bent-up dipper. The Big Dipper also goes by other names. The English see it as a plow. Egyptions see it as an axlike tool called an adze. The imaginative ancient Greeks saw the Big Dipper as the tail and hind part of a great bear and the Little Dipper as a small bear.

On some clear, moonless night, try looking for these two bears in the northern sky. If the Greeks could see bears, we should be able to also. Besides, the official name for these two constellations is Latin for Great Bear and Little Bear: Ursa Major and Ursa Minor.

Orion

Winter is a great time for looking at stars. Nightfall comes early. The air is clearer, because colder air can hold less moisture. There are twice as many bright stars in the winter sky as at any other time of year. And the winter sky has a spectacular constellation that is not only large and made up of bright stars, but actually looks like what it is supposed to be, a giant hunter.

Orion, the Hunter (a character in Greek mythology who was placed among the stars when he died), rises in the east about midnight in September. As the months go by it rises earlier and earlier. By November Orion is in the southeastern part of the evening sky. By January it is already high in the southern sky at sunset. And come April, the Hunter is setting in the west just after the sun.

The key to finding Orion is his belt, which is made up of three bright stars equally spaced in a row. A short sword dangles from his belt. His shoulders and knees are clearly marked with bright stars. There is a star for his head, but it isn't in quite the right position. With a little imagination, some faint stars form his right arm, holding up a club, and his left arm, holding out a lion skin for a shield.

Don't bother to look for the Hunter's feet, for where his feet should be is the constellation Lepus, the Hare. Orion's dog is easy to find, though. Follow the line of Orion's belt to the east and you quickly come to the brightest star in the heavens. This is Sirius, the Dog Star. Sirius and several of the stars around and below it form the constellation Canis Major, the Great Dog.

If you have binoculars, the middle star of Orion's sword deserves closer inspection. Fuzzy-looking to the naked eye, the star still looks fuzzy through binoculars, because what you see is really starlight shining off a tremendous cloud of dust called the Great Nebula of Orion.

Also take a closer look at the star that marks Orion's right shoulder. It is a giant red star with the Arabic-derived name of Betelgeuse (pronounced like "beetle juice," except with a *z* sound at the end instead of an *s* sound). You'll notice a definite orange-red tinge to Betelgeuse, especially when you compare its light with the blue-white light coming from the star Rigel, located kitty-cornered at Orion's left knee.

The Pleiades

For thousands of years people have used the Pleiades as a test for good eyesight. On any clear, moonless winter evening, especially in January and February, you can take the test yourself and continue an ancient tradition. The only problem is that nowadays the Pleiades might test how clear your air is or how much interference the lights of the nearest town cause rather than how good your eyesight is.

Astronomers say the Pleiades (pronounced PLEE-a-deez) are a tight cluster of stars, but classical mythology has it that they are the seven daughters of Atlas and Pleione. Long ago the mighty hunter Orion fell in love with these seven beautiful sisters. As Orion pursued them, the sisters needed help, so the gods turned them into doves and placed them among the stars. To this day, Orion still pursues them across the night sky.

To find the Pleiades, first locate Orion (see page 110). Then, remembering that Orion is still chasing the seven sisters and that constellations, like the sun, move from east to west, look up and to the west of Orion and you'll see them. The Pleiades are in the shoulder of Taurus, the Bull, but about all that can be easily recognized of the Bull is the face and V of the horns looming over Orion.

How many Pleiades can you see? If you can see all seven of the sisters, you have good eyesight. If you can only see six, you still have pretty good eyesight. One form of the legend says that there are only six of Atlas's daughters still up there because one of them left her place in the sky so she wouldn't have to see the fall of Troy. If you see more than seven, which is possible with the naked eye, you must have very good eyesight and very clear skies. It's not fair to use binoculars during the test, but if you do, you should be able to see thirty or more stars in the Pleiades cluster.

The Milky Way

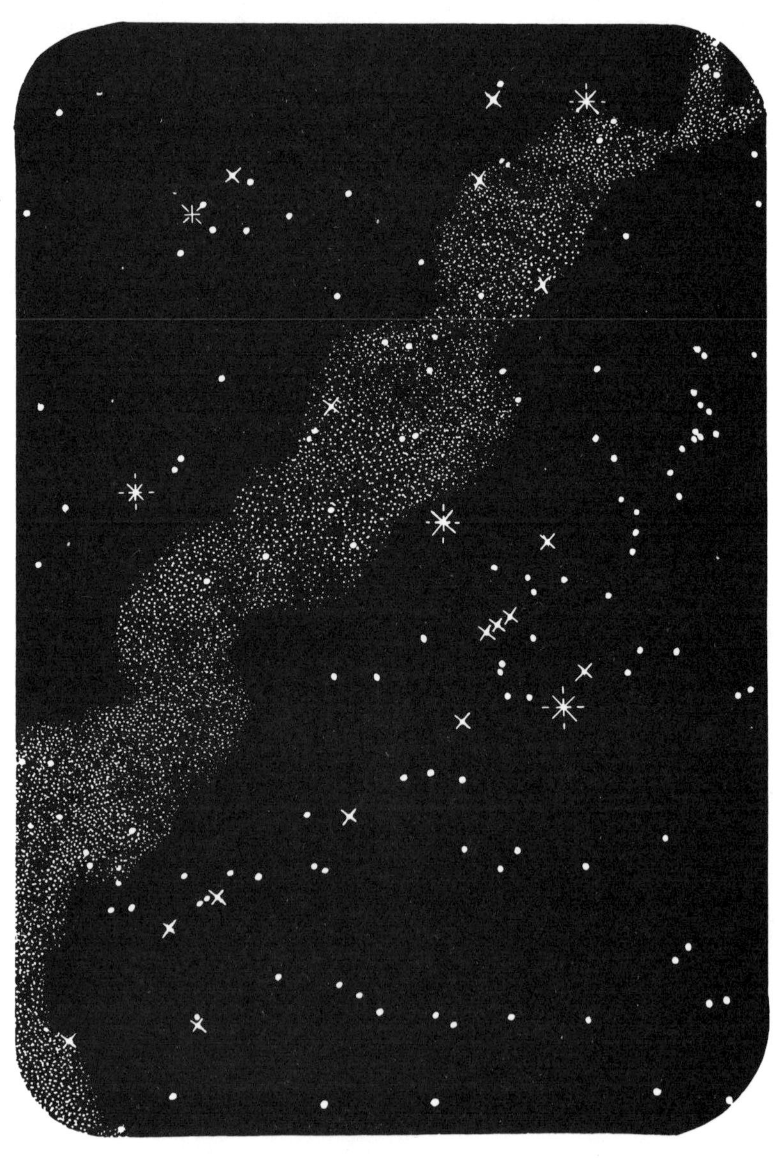

The Milky Way is a glowing band of soft starlight meandering sometimes from north to south and sometimes from east to west. You can see the Milky Way any night of the year so long as the sky is clear, the moon isn't out, and you are far from city lights. These three conditions are crucial, because most of the Milky Way is very faint and hard to see against any other light.

The one third to one half trillion stars in the Milky Way Galaxy form a flat, pinwheel-shaped spiral with a bulge in the center. All the stars you see, including the sun, are part of the Milky Way Galaxy. The Milky Way we see in the night sky is the more distant part of our flat galaxy.

The best time to see the Milky Way is in evenings in late summer. The Milky Way is at its widest and brightest then, because you are looking toward the center of the galaxy. In late summer evenings you'll see the Milky Way passing directly overhead, running from the northeast to the southwest and covering over 10 percent of the sky (its direction is determined by the time of night and the time of year). The Milky Way stays high in the sky in the evening all fall and winter. During winter you can see the Milky Way passing right through Orion's raised right arm (see page 110).

A number of different cultures have legends that portray the Milky Way as a pathway for gods or the souls of the dead. In Roman legends the milky light came from an overflow of milk for the baby Mercury. The Milky Way has also been likened to rivers and given names like the Silver River, the Celestial River, and the River of Light. Legends in some lands, such as Armenia, say that the Milky Way is wheat, rice, ashes, or straw dropped or tossed by various heroic figures. My favorite Milky Way legend, from the Buriats of Mongolia, says the Milky Way is the seam where the sky was sewn together.

Many legends grew up about it, but the true nature of the Milky Way was known way before modern times. In the fifth century B.C. the Greek philosopher Democritus figured out that the Milky Way's misty glow was the united light of many far-off stars.

Look at the Milky Way with the naked eye or binoculars and you'll notice that it is not an even band of light. There are bright areas and dark areas and even a part where the Milky Way splits in two, called the Great Rift. Most of the dark areas, including the Great Rift, are caused by massive clouds of interstellar gas and dust blocking out the light of millions of distant stars.

Venus

Venus is the easiest planet to identify, because it is always seen close to the sun either just before sunrise or just after sunset and it is very, very bright. Far and away the brightest "star" in the sky, and therefore aptly named after the Roman goddess of beauty and love, when Venus is visible in the evening it will be the first star you'll be able to see, even though it will be close to the setting sun, where the sky is lightest.

Venus watchers shouldn't wait for the sky to get completely dark, because Venus will frequently have set by then. Venus is always seen fairly close to the sun because its orbit is closer to the sun than Earth's orbit. Venus never appears more than about 45 degrees away from the sun, which means that at sunset you'll never see Venus higher than halfway up to the zenith. At its highest, Venus will rise in the morning or set in the evening only three hours before or after the sun.

Once you get to know Venus, you'll be surprised how often you'll see it. Venus's brightness varies, but at its brightest this planet sends us twelve times more light than the Dog Star, Sirius, which is our brightest true star.

Look at Venus with binoculars when it is safely away from the sun's glare and you'll notice that it isn't round like stars. Instead, Venus goes through phases like the moon, which is why its brightness varies. You'll frequently see a half-Venus shining like a half-moon. But you'll never see a full Venus, because when Venus is full it is too close to the sun to be seen.

The following table can tell you when and where to look for Venus. Since Venus doesn't pass instantly from being hidden in the sun's glare to being visible, the beginning and ending dates for each of Venus's runs can only be approximate.

Visible in the Evening	*Visible in the Morning*
mid-Mar. '86–late Oct. '86	mid-Nov. '86–late June '87
late Oct. '87–late May '88	late June '88–early Feb. '89
early June '89–early Jan. '90	late Jan. '90–mid-Sept. '90
late Dec. '90–mid-Aug. '91	early Sept. '91–late April '92
late July '92–late Mar. '93	mid-April '93–early Dec. '93
mid-Mar. '94–late Oct. '94	mid-Nov. '94–late June '95

The dates Venus is visible happen to follow an eight-year cycle (minus two or three days). So to find out when and where Venus is visible in the year 1996 or 2004, for example, just refer back to the proper times for 1988.

BIRDS

Barred Owls and
Great Horned Owls

You may hear an owl calling in the night anytime of year and just about anywhere, from forests to open country, from wilderness swamps to suburban areas. The best time to hear owls is on a still evening in early spring, because owls tend to call more often during their nesting season. Try listening near some woods after the sun has set but before the sky has completely darkened, and you may hear the deep hooting of an owl.

You'll find it surprisingly easy to identify owl calls, because North America only has two common owls that routinely call with deep "hoos." If the call sounds like the standard deep owl hoot you might hear in a movie, you are most likely listening to a barred owl or a great horned owl. To tell them apart, listen carefully to the last hoot of the call. The last hoot of a great horned owl call sounds like all the other hoots and just fades away: "hoo, hoohoo, hoohoo, hoo." The last hoot of a barred owl call sounds quite different from the other hoots. Its last hoot ends with a croaky "aw" that descends in pitch: "hoohoo, hoohoo, hoohoo, hoohooaw." It is almost as if the barred owl ends by saying "you-all."

Great horned owl "horns" are tufts of feathers above their ears. Great horned owls are larger and more powerful than our largest hawk. Rabbits are their main food, though they do eat all sorts of birds and animals, including skunks. Their calls are deep and resonant. On a still night you can hear a great horned owl calling from quite a distance. Most great horned owl calls are a series of four to eight hoots.

Barred owls got their name from their streaked or barred breasts. They are smaller than great horned owls, and typically eat mice, rats, and frogs. Barred owl calls usually have eight hoots with the croaky "aw" at the end.

What about the seventeen other species of owl in North America? Most of them have calls that are better described as whistles, wails, hisses, or sneezy barks than hoots. A couple of owls, like the great gray owl, will hoot, but they are so rare that you aren't likely to hear them. And a couple of owls, like the long-eared owl, occasionally will hoot but usually are silent. The robin-size western screech owl, however, does call with a fast series of very short hoots; but these are much shorter and quicker than the classic deep hoot you usually associate with owls.

Nighthawks

Nighthawks have long, pointed wings like falcons, but they are not hawks. Nighthawks fly at night, but they are not purely nocturnal. You can often see them flying in the afternoon and evening. Nighthawks are gray-brown birds with almost a two-foot wingspan. You can easily identify them by the white band that cuts across their long, pointed wings.

Strange as it seems, the best place to see nighthawks is in small towns, suburbs, and cities, though you can also see nighthawks in open or partially wooded country. Look for them in late afternoon and evening flying at treetop height and higher. You'll see sprightly nighthawks darting after flying insects, which they catch in their huge mouths.

Nighthawks lay their eggs right on the ground in rocky or gravelly areas. When people started building flat roofs covered with tar and gravel, nighthawks moved into towns. Thanks to this architectural style, nighthawks have gravelly places to nest that are free from most predators, and urban people have the treat of seeing nighthawks swooping in and out of the glow of streetlights.

North America has two species of nighthawk, the common nighthawk and the lesser nighthawk. The common nighthawk lives throughout most of the United States and southern Canada, while the lesser nighthawk is limited to the Southwest, from central California to southern Texas. The lesser is only a little smaller than the common. Birders in the Southwest can tell the lesser from the common nighthawk by the lesser's habit of flying low to the ground rather than at treetop height and higher.

They can also be told apart by their calls. The lesser nighthawk gives a sweet, trilling call, while the call of the common nighthawk might be described as a raspy buzz or a nasal "peent." This is not a very musical call, but it's easy to remember once you've heard it. On warm summer evenings you can quickly recognize the presence of common nighthawks simply by their calls.

When you watch nighthawks in late spring and early summer, you may see the male nighthawk perform an aerial display that is part of his mating ritual. He will fly quite high and then dive down to earth with his wings folded. As the male opens his wings to pull up from the dive, the long flight feathers on the wings vibrate to create a loud noise that sounds much like the twang of a fat rubber band.

Chimney Swifts

Chimney swifts nest and roost in chimneys, so the best place to see chimney swifts is where chimneys are common—in villages, towns, and cities. Look for these dark, swallowlike birds in the half-light of sunset on a warm summer evening. You'll often see them in chittering flocks flying around at treetop height and higher. Keep a close watch on individuals that fly near chimneys, to see if the bird disappears down a chimney. Seeing the swallowlike birds fly into or out of a chimney will give you a positive chimney swift identification.

Swifts were given their name because they are among the world's fastest birds. Their stiff wings curving back in a crescent are well adapted to continuous high-speed flight. Chimney swifts have very short tails that help prop them up when they roost on vertical surfaces. Short tails and stiff wings have given chimney swifts a flying style that can help you recognize them. Watch chimney swifts you've positively identified and you'll notice that they fly with jerky, uneven beating of their wings.

During the day you can tell chimney swifts and swallows apart by how they fly. Swallows generally beat their wings evenly and turn by twisting their tails. Chimney swifts use their little tails, but to make sharp turns chimney swifts must beat one wing harder than the other. They are also sometimes mistaken for bats, which also flap their wings unevenly. In the evening you can easily tell chimney swifts from bats by how their wings are connected to their bodies. Chimney swifts have narrow wings connected to cigar-shaped bodies. Bats have wide wings attached to the whole length of their bodies.

Chimney swifts are found in the United States and southern Canada from Colorado east. They are the only swift found in most of this area. In the days before chimneys, chimney swifts nested in hollow trees. The three species of swift in western North America are not as common or as urbanized as chimney swifts and still nest in hollow trees and rock crevices.

Chimney swift saliva is very sticky, which must be helpful to birds that catch flying insects in their mouths. Chimney swifts build nests by gluing twigs to the inside of chimneys with their sticky saliva. In Asia there is a swift that builds its entire nest out of its thick saliva. These are the nests people eat in Oriental bird's nest soup.

Mourning Doves

In the still dawn of an early spring morning the soft cooing of a mourning dove carries a long distance. Mourning doves got their names because their call sounds sad and mournful to people. But to other mourning doves the calls may sound proud and robust. Mourning dove calls are a series of four soft, slow, low-pitched coos, with the first coo rising in pitch and the last three all at the same, lower pitch. The coos are so low-pitched that they may remind you of owl hoots.

The best time to hear mourning doves cooing is any time of day in the early spring when males are marking off their territories. The males usually call from open perches, such as power lines and dead branches of trees. One of the easiest ways to see a mourning dove is to follow the sound of the spring cooing.

The key to identifying mourning doves when you see them is that their long tails taper to a point. This is quite easy to see when they fly. Mourning doves are gray-brown birds with small heads and skinny necks. Look closely and you'll notice a subtle pink shading to some of the brown. Mourning doves are smaller and slimmer than their cousin the common pigeon. You can easily learn to identify mourning doves by their pointed tails, by their "mournful" cooing, or by the whistle of their wings. Mourning doves are fast and strong fliers. When they fly you can hear a musical whistle of air passing through their beating wings.

The best places to see mourning doves are towns and farmland. Mourning doves like partly open country where they have open ground to forage on and trees to nest and roost in. Look for mourning doves in city parks, in grassy fields, at roadsides, and at bird feeders. Mourning doves feeding in a farm field generally don't hurt the crop, because they forage for seeds on the ground. They may even help farm crops by eating lots of weed seeds.

American Kestrels

Most of our falcons are magnificent birds that inhabit wild country. The American kestrel, however, is better described as a dainty dweller in farms and towns. You can find kestrels, also called sparrow hawks, in open and partially open country as well as in suburbs and city parks. Look for them on power lines and outer branches of trees that give them a good view of a grassy area. Kestrels hunt most actively in the morning and late afternoon, but they will sit out in the open even when they are not hunting.

Measuring in at about the size of a robin or a little larger, kestrels are the smallest of the hawks and falcons. You can identify them by the typical hooked bill of a hawk and their rusty, or rufous, backs. No other small hawk or falcon has a rufous back. Kestrels are one of the few species of eagle, hawk, or falcon that have males and females colored differently and young colored the same as adults. You can tell male and female kestrels apart by the color of their wings. The male's wings are blue-gray, while females have streaked, rufous wings the same color as their backs and tails.

In the Southeast the best time to see American kestrels is during the winter. Many kestrels from Canada and the northern United States go south for the winter, creating a noticeable population increase. In the rest of North America kestrels are usually easiest to see in late summer after the young have left their nests.

Watch a kestrel hunting from a perch and you may see it fly off, hover in midair for several seconds, then dive headfirst on partly folded wings. A kestrel catches prey with its feet, then does the killing with its beak. Usually the kestrel will fly back to its perch to eat. American kestrels eat just about any small creatures that are available, such as mice, birds, and lizards. During the summer this dainty falcon often feeds primarily on grasshoppers and crickets. Kestrels in cities rely mainly on house sparrows. American kestrels adapt to cities partly because they aren't overly afraid of people, and because the small prey they live on exists there.

Turkey Vultures

With six-foot wingspans, turkey vultures are masters of effort-less soaring. They soar around and around for hours, looking for food. Therefore, the easiest way to find a turkey vulture is to simply look up occasionally when you are in the country. Turkey vultures can quickly be told from other large soaring birds by either their color or the way they hold their wings.

The body and lead part of turkey vulture wings are dark gray while all the long flight feathers in the wings are light gray. No other large soaring bird has this simple, two-tone coloring. When the sun is high and shining through the bird's wings, the two-tone-gray color pattern is very clear and obvious. Sometimes on cloudy days, though, all soaring birds appear as dark silhouettes. When this happens, you can recognize turkey vultures by the way they hold their wings.

Turkey vultures routinely fly with their wings tipped slightly up, forming a shallow V. Most other soaring birds keep their wings horizontal. The shallow V increases the vulture's aerody-namic stability. You'll find this distinctive V easy to see when the birds are flying somewhat toward you.

Turkey vultures have featherless, red heads that you'd think would be an obvious identification feature. But they aren't. Turkey vultures usually fly too high to let you see clearly what their heads are like. About all you'll be able to tell about their heads is that without feathers (a feature that is thought to keep them cleaner) they appear smaller than you'd ordinarily expect for birds with a six-foot wingspan.

Our culture seems to look down on vultures and other scav-engers. Vultures, however, are an important part of our ecosys-tem. Their quick and efficient disposal of dead animals helps keep diseases down. Some other cultures seem to appreciate the vul-ture's special talents more than we do. For instance, instead of burying or cremating their dead, members of the Zoroastrian faith (originally Persian but now mostly Indian) traditionally placed corpses on special towers where vultures could land and dispose of the flesh.

Red-tailed Hawks

The exact coloring of a red-tailed hawk varies across the country. The upper side of the wings and backs of these large, powerful birds are always brown, usually a dark, tree-trunk brown, but their breasts vary from fairly dark to mostly white. You can, however, quickly and positively identify red-tailed hawks by their red tails.

This may sound straightforward, but identifying red-tailed hawks isn't quite as easy as it sounds, because only adult red-tailed hawks have red tails and then only the upper side of their tails are red. The underside of their tail feathers are a pale white that lets only a little of the reddish color show through. To complicate matters even more, their tails aren't a true red. Red-tailed hawk tails are more of a rusty, burnt-orange color, like a robin's breast.

Red-tailed hawks are seen throughout most of the country, usually conspicuously sitting in a tree at the edge of a field or soaring around and around overhead. Soaring hawks turn by twisting and tipping their tails, so if you keep watching a soaring hawk, pretty soon you'll get a glance at the top of its tail—perhaps only a fleeting glance, but that is usually enough to see if the whole top of the tail seems to be the color of a robin's breast. If it is, then the bird is a red-tailed hawk one year old or older.

Red-tailed hawks typically hunt ground animals, such as mice, rats, and rabbits, in open fields. The red-tails' four-foot wingspan makes it easy for them to soar for hours, but also makes them fairly slow when it comes to level, flapping flight. To gain hunting speed, red-tails like to fly down from high vantage points, so look for red-tails high in trees or on the top of poles near the edge of a field. When in a tree, they'll sit conspicuously at the edge of the tree, scanning the ground for prey.

Red-tails nest in woods, but their long, broad wings make maneuvering too difficult for them to hunt very successfully there. The combination of needing trees for nesting and fields for hunting means that red-tailed hawks are fairly rare in completely forested areas and in completely treeless areas. In the rest of North America red-tails are the most common and widespread of our large hawks.

Crows, Hawks, and Ravens

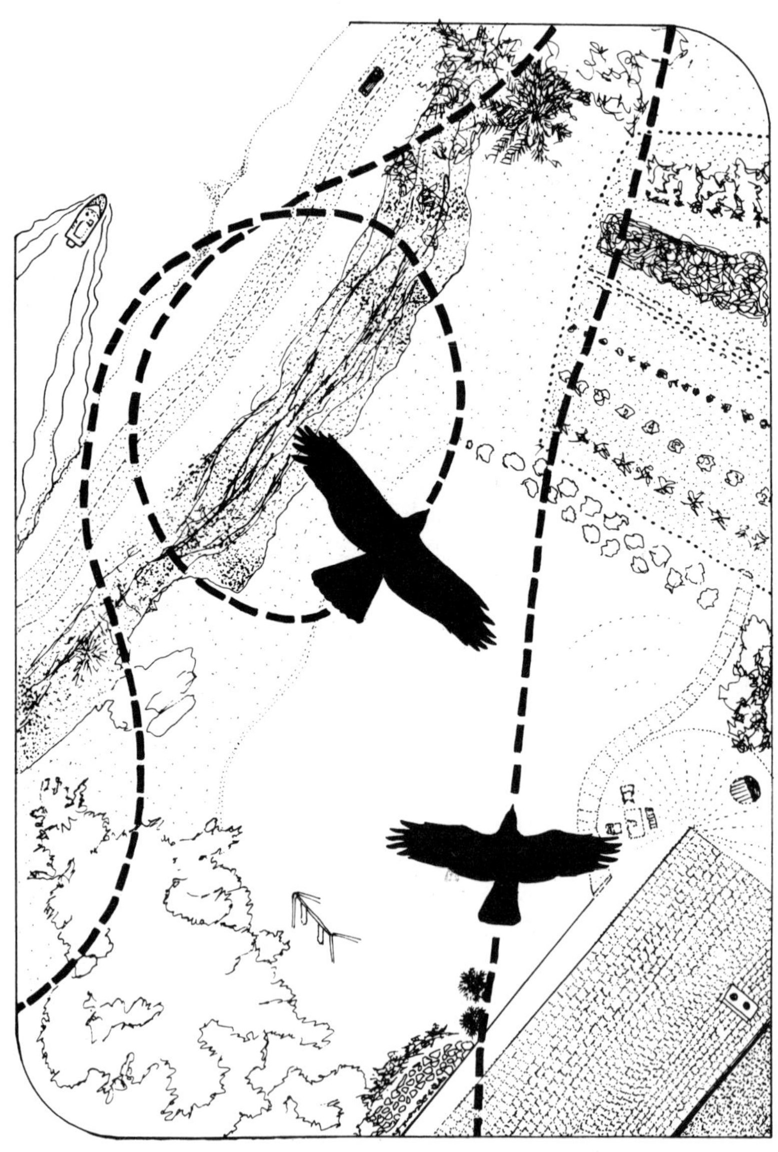

You're out for a hike and you see a large bird flying in the distance. The bird is so far away that you can't tell what color it is. All you can see is a silhouette. From the silhouette you can tell that the bird doesn't have the long neck of a duck or a goose, the pointed wings of a sea gull, or the long bill and trailing legs of a heron. And it is not a hawk, because they soar and this bird is flapping its wings. Chances are, then, that it's a crow.

Crows will seldom soar for more than a few seconds. They usually flap their wings continuously. Flying this way takes more energy than soaring but allows crows to fly in straight lines. And crows usually do fly in straight lines when going from one area to another. Over hills and dales, they'll fly straight, taking the most direct route to where they are going. That is why when you talk about how far away some place is in a straight line, you say it is so many miles away "as the crow flies."

You won't hear "as the hawk flies," because most of the hawks you'll see are types that soar whenever they can. They flap their wings when necessary, but most often soar on wind currents. This soaring style of flight involves a lot of circling and taking more roundabout routes. It also allows hawks to stay in the air for a long time without large expenditures of energy. Hawks can put this time to good use by hunting as they soar.

This flapping/soaring identification technique also works fairly well for telling crows and ravens apart. Ravens frequently soar like hawks. And even in ordinary flight, ravens will glide between periods of flapping. The best way to tell ravens and crows apart, though, is by the shape of their tails. Crow tails are fairly flat across the end. Raven tail feathers are longer in the center, making for a wedge-shaped tail.

Crows and hawks can be found throughout most of the United States and southern Canada. Crows will eat just about anything and are found in a wide variety of habitats, ranging from forests to farmland and from tidewater marshes to deserts. You may think that black is a funny color for a day bird, that it would make crows an easier target for predators. But the crow's biggest enemy is the great horned owl, which catches crows at night while they are roosting.

Belted Kingfishers

You can find kingfishers wherever there is water, from creeks to rivers, ponds, lakes, and even ocean coasts. Sometimes you'll see kingfishers flying over the water, perhaps stopping to hover and dive for a fish, but most of the time you'll see a belted kingfisher sitting alone on a perch that hangs out over the water. Dead trees are often the preferred perches, because they don't have leaves obstructing the view.

Belted kingfishers are almost as large as pigeons and can easily be recognized by their large heads with ragged double crests on top and huge bills. Belted kingfishers are a pretty but not showy blue-gray color above and white below. The belt for which they are named is a band of blue-gray that cuts across their white breasts about even with their shoulders. Above the blue-gray belt they have a white collar that runs almost all the way round their necks.

If you get a good view of a belted kingfisher, you'll also be able to tell which sex it is. Belted kingfishers are one of the few species of North American birds with females more colorful than males. This color takes the form of an additional breast band, which is rusty-brown. The male doesn't have any rusty-brown coloring.

Canoeists probably see more kingfishers than anybody else. And each kingfisher sighting is special. Kingfishers are beautiful, fairly good size, and interesting to watch. Even a kingfisher scared up by an approaching canoe will often fly away with style by swooping down off its high perch, flying low to the water, and then swooping almost vertically back up to a new perch. Kingfishers commonly have a circuit of favorite fishing perches that they visit during the day.

From these perches a kingfisher scans the water, looking for signs of small fish, salamanders, crayfish, tadpoles, frogs, and even insects. Spotting a fish, it launches off its perch and dives headfirst after the fish, going completely underwater if necessary. With strong wings, it flies up out of the water and returns to its perch. Almost all fish-eating birds are careful to swallow the fish headfirst so the fins and scales go down smoothly. The kingfisher is no exception. With the fish held firmly in its beak, the kingfisher stuns the fish by banging its head against the perch. Then, with a deft flick of its head, the kingfisher tosses the fish into the air and swallows it head first.

American Coots

Most people assume that American coots are ducks until they see a coot's feet. Coots don't have webbing between their toes, as do ducks. Instead, they have little lobelike flaps on their toes to help them swim. Coots' toes are very long, allowing the coot to walk on top of and forage among marsh plants, so look for coots on lakes and rivers that have some marches along a shore. Coots seem to have less fear of people than most waterfowl and will often let you get close enough to get a good view. The best time to look for American coots is in the fall, when they gather in flocks. Sometimes you'll see hundreds of coots in a single flock.

You can also identify an American coot by its dark gray coloring and white bill. American coots have a little white on their rumps, but basically they are all gray. The white of the beak is quite easy to see, as it contrasts with the very dark gray of the face feathers. Coots have very short, almost useless tails. American coots fly with their legs and long toes hanging back, and actually steer more with their feet than their tails.

American coots are fairly good fliers once they are airborne. The problem is getting airborne. To take off, coots run along on top of the water, flapping their wings like mad. A whole flock of coots taking off and splashing with each step as they run along on the water is a treat for your ears as well as your eyes. Sometimes coots hurrying away from danger will run on the water with wings flapping, only to settle down again when they reach safety. Safety for coots may mean open water or it may mean hiding in a marsh. In the spring American coots build floating nests, which they anchor and conceal among the marsh plants.

Coots eat all sorts of aquatic plants as well as small fish and insects. They search for food in many different ways. You'll see them swimming and walking in shallow water, going around and over marsh plants. And you'll see them dive completely underwater. Coots have a quick, headfirst surface dive that any human swimmer would envy.

Sometimes you'll see coots eating grass and seeds on land. The coot's long toes help make it fairly awkward out of water. Watch a coot clumsily foraging for food on land and perhaps you'll understand why the word "coot" is used to mean a harmless, silly old man.

Mallards

Often the best places to see mallard ducks are ponds in city parks. Look for mallards during the summer in the North and during the winter where ponds don't freeze over. In city ponds they often become tame enough to take bread scraps out of your hands. In the wild where they are hunted each fall, you will have a hard time getting close to a mallard.

The most obvious feature of the male mallard is its green head. Several other species of duck also have green heads, so to positively identify male mallards you should notice that they have a white neck band below the green and that their heads don't have crests. Males are mostly gray with brown chests. Positively identifying female and juvenile mallards is difficult, because their blotchy brown coloring is about the same as that of several other species of duck. When a flock has only male mallards and mottled brown ducks, it's usually safe to assume the brown ducks are female or juvenile mallards.

Mallards feed mostly in the shallow water of ponds, lakes, and rivers. You'll often see them tip up, with their heads down underwater and their tails up in the air. They will mainly be eating the seeds of plants such as coontail, bulrushes, and sedges, which they find on the bottom of the pond. Mallards also eat small animals such as snails and tiny fish. Sometimes you'll see mallards foraging for seeds on land as well.

Look closely at a flock of mallards and you'll notice that only the females quack. The males give softer, reedier calls. The female's quack sounds like the stereotypical domestic duck quack because mallards are the ancestors of most of our domestic ducks. Sometimes you'll see strange-colored ducks in city ponds, the result of mallards breeding with white domestic ducks. Mallards usually build their nests on land, hidden in tall grass or under bushes. The ducklings are ready to follow their mother into the pond shortly after hatching.

Canada Geese

Look for Canada geese around ponds, marshes, lakes, and rivers. You'll often see semidomesticated Canada geese in ponds by golf courses or in city parks. Most Canada geese nest in Alaska and northern Canada. In the fall they fly south to winter on water that doesn't freeze over. In most of the United States and southern Canada the best time to look for wild Canada geese is during migration in the fall and early spring.

Canada geese vary considerably in size and length of neck. Some Canada geese are only a little larger than ducks, while others are almost twice that size and weigh three times as much. Fortunately, all Canada geese are colored about the same, so no matter what size, you can identify Canada geese by their black necks and heads with white cheek patches that go from under their chins up to about where their ears would be. Except for the head and neck, these birds are mostly brownish gray.

Geese are grazers and eat grass leaves, bulrushes, and many other plants. They also love corn. When migrating Canada geese stop to rest and feed, they often land in grassy areas to graze or in farm fields to scour for grain. Canada geese also congregate in marshes, where they graze both above and below the water surface. You'll find that marshes provide especially interesting bird-watching in the spring, when waterfowl are migrating, blackbirds are marking their territories with song, and mosquitoes aren't out yet.

Migrating Vs of Canada geese fly over almost all of North America. To see a flock next spring or fall, you should keep your ears cocked for the geese's musical honking whenever you are outside, day or night. Once you hear their honking, search the sky for the V.

Canada geese aren't the only birds to fly in V formation, but they are the most common. Canada geese almost always migrate in formation, usually in a V but sometimes in a W or just a slanted line. When a bird flaps its wings, a stream of air swirls up around the tip of each wing, creating vertical eddies, or spinning loops of air. Geese flying in formation apparently arrange themselves so part of one wing extends into the upward swirling part of an eddy created by the bird in front. This makes flying somewhat easier on all but the bird in the lead.

Killdeers

It's late spring and you are walking across a golf course. Suddenly you notice a crippled bird about the size of a robin. The poor bird is hobbling along, dragging one wing on the ground. You can tell by its peeping and the way it flutters its broken wing that the bird is in distress. Should you try to help it or not? As you hesitate, it flops on the ground, broken wing stretched out farther, peeping even louder. You decide to help. As you walk toward it, the bird hobbles away. You run to catch up and the bird flies a short distance, lands, and again displays its broken wing. After you have followed it for about fifty yards, the broken wing miraculously heals and the bird flies off, calling loudly in a high voice.

The noisy actor or actress you just followed was a killdeer. Killdeers and a few other ground-nesting birds have found the crippled-bird act to be a very effective way to lure predators away from the killdeers' eggs or young. And bird watchers have found it an effective way to get a good close look at a killdeer. (The bird got its name because its loud call sounded to some like the words "kill deer.")

Killdeers are about the size of a robin, with mostly brown coloring above and white below. You can identify adult killdeers by the two black bands across their breasts. Killdeers also have orange rumps, which aren't very noticeable until they put on their broken-wing routine. These noisy birds may repeat their call over and over again as they fly.

You'll find killdeers on plowed fields, lawns, beaches, and other types of open country. They are fast, graceful fliers and can run as fast as any shore bird you might see on a beach. Typically, you'll see killdeers standing or running along on the ground, looking for insects.

Killdeers are around just about anytime the ground isn't frozen. But if you want to watch a killdeer do its crippled-bird act, you should look for them in late spring or early summer, when they have eggs or chicks to protect. Killdeer eggs look like four identically speckled rocks set together in a little dip in the ground. The most common nesting sites are open places with enough gravel to make the eggs hard to spot. Killdeer chicks can run along and forage for food behind their parents almost immediately after hatching. The young chicks can't fly, though, so they still need their parents to lead predators away from them with the broken-wing routine.

Brown-headed Cowbirds

If you peek into bird nests whenever you get a chance, sooner or later you're bound to find a nest with two different-colored or different-size eggs. This means that a cowbird has been there and laid its eggs in another bird's nest. Brown-headed cowbird eggs are white with brown specks. An easier rule of thumb to remember when identifying whose eggs are whose is: Cowbird eggs are usually larger.

Cowbird eggs have a short incubation time, so they usually hatch before the host bird's eggs. Couple earlier hatching with larger size and the cowbird babies can easily dominate their nests, thereby ensuring their own survival in hard times over the host bird's own babies.

Cowbirds are small blackbirds, but larger than the vireos, warblers, and sparrows whose nests they prefer. The male brown-headed cowbird is very easy to identify. He is all black except for his brown head and neck. Female brown-headed cowbirds are a little trickier to identify. They are gray, and are usually identified by being the same size and shape as nearby male cowbirds.

Cowbirds got their name from their habit of following herds of cows so they could eat the insects and seeds exposed by the trample of the cows' hooves. In the days when the cowbird evolved this feeding technique, the "cows" they followed were the vast, roaming herds of buffalo, which did not stay in one place long enough for cowbirds to raise their own young. To be able to follow their source of food, cowbirds developed the habit of having other birds raise their young.

Brown-headed cowbirds are prairie birds. Though they range into open woodlands, brown-headed cowbirds won't go far into a forest to find a host nest. When forests were vast, the woodland vireos had many nests cowbirds could never reach. But now forests are smaller and farms, cows, and cowbirds are all over the country, making for fewer safe vireo nests. Therefore, cowbirds can now have a much more adverse effect on host bird populations than ever before.

You no longer need to go to a herd of cows to find cowbirds. Although they are still common around cattle, you can also find brown-headed cowbirds on other agricultural land as well as in suburban areas, golf courses, and city parks. You'll often see cowbirds in flocks with other blackbirds.

Horned Larks

You can positively identify horned larks by their black "sideburns" and other distinctive facial markings. At first glance you'll notice the vertical black streaks across their yellow-and-white faces. Looking through binoculars, you'll see that horned larks have three separate black markings: a black mustache and sideburns, a black forehead and horns, and a large black mark across the top of the breast. These dark marks show up distinctly against the yellow-and-white face and white breast. Yet horned larks aren't flashy birds. They are only a little larger than sparrows, and their backs and wings are a drab, streaky brown.

Horned larks are ground birds that prefer large, open fields with sparse or short vegetation. You can find horned larks in golf courses, pastures, plowed fields, prairies, deserts, tundra, and beaches. Look for them walking or running on the ground and flying low to the ground. Horned larks occasionally land on fences and other low objects, but you'll almost never see a horned lark in a tree.

In the southern half of the United States and along the coasts the best time of year to see horned larks is during the fall and winter, when local horned lark flocks have swollen with larks migrating down from the arctic tundra. In southern Canada and the northern United States the best time of year to see horned larks is in the early spring, when some but not all of the snow has melted. This is when migrating horned larks are heading back to the Arctic and local horned larks are staking out nesting territories.

Horned larks are one of the earliest nesting songbirds, though their squeaky song is nowhere near as beautiful as the famous melodies of European larks. They build their nests right on the bare ground, and so early that the nests are often buried by spring snow. As grass and weeds start to grow thick, horned larks tend to leave the area in search of more barren ground.

How do small, ground-dwelling birds manage in winter? Well, horned larks look for food on shores and in windswept areas where snow has been washed or blown away. They pluck seeds from grasses and weeds that stick out above the snow. And they pick bits of undigested grain out of animal manure.

Meadowlarks

Meadowlarks can be easily and positively identified by the bold black V crossing their bright yellow breasts. The bright yellow of their breasts goes all the way up to their bills, with only the black V interrupting the yellow. Meadowlarks' heads, backs, and wings are colored with drab brown streaks that blend in with dead grass and weeds. The outside tail feathers are white, while the center of the tail is more of the same brown, dead-grass color. Meadowlarks are about the size of robins but have shorter tails.

With a name like meadowlark, you should know right away that they are birds of meadows, prairies, and fields and that they probably sing wonderfully. The best time to look for meadowlarks is in spring, when the males are staking out their territories. Male meadowlarks mark and defend their territories by loud singing from the tops of weeds, fence posts, and electric wires. Look for meadowlarks in any large meadow or grassy field. They are ground birds, so you will usually see them close to the ground on weeds and fences.

Flying meadowlarks have an interesting habit of flapping their wings quickly and then holding them stiff and gliding. They often glide in low for a long time before they finally land. If they land on a tall weed, they are easy to see. But if they land on the ground among tall grass, they practically disappear. Meadowlarks spend a lot of time walking around on the ground among the grasses and weeds, looking for insects. They eat a wide variety of insects as well as some grain and weed seeds. They nest right on the ground, but you'll usually find a meadowlark nest only by accident. They hide their domed-over grass nests in clumps of weeds, and meadowlarks heading to their nests always land some distance away and walk to the nest.

We have two species of meadowlark in North America, the eastern meadowlark and the western meadowlark. These two species look practically identical, so in the central part of the country, where both species live, avid bird watchers tell them apart by their songs. To accurately tell the difference between the eastern's clear, slurred whistle and the western's flutelike gurgling you'll need to hear them both, either live or recorded.

American Goldfinches

In spring and summer male American goldfinches are conspicuously decked out in bright lemon yellow with contrasting black caps, wings, and tails. There are no other sparrow-size birds with similar coloring. Females and males during the fall and winter camouflage themselves with drab olive-yellow coloring that makes them harder to see and trickier to identify. A flock of a dozen American goldfinches can fly by in late fall with hardly anyone bothering to look at them. But if one bright-yellow-and-jet-black male flies by in late spring, heads will turn to admire his beauty.

Look for American goldfinches in weedy fields, roadsides, brushy thickets, trees at the edge of fields, or open woodlands with lots of weeds growing between the trees. The best time to see American goldfinches is in spring, when the males have changed into their bright colors yet the goldfinches are still in their winter flocks. American goldfinches stay in winter flocks much later than other birds, because in most places they don't start nesting until July.

American goldfinches will eat all sorts of seeds, from elm seed to dandelion seed, but their favorite is thistle seed. They don't start nesting until the thistle seed crop is ready to feed their young. They even line their nests with thistle down.

When you watch American goldfinches fly, you'll notice that they fly in a distinctly roller-coaster fashion, similar to the way woodpeckers fly. This undulating flight pattern can help you identify the small American goldfinches on the wing. Listen closely as you watch them fly and you'll also notice that they often sing a special little flight call on the upswing part of their flight.

You can attract American goldfinches to backyard bird feeders with sunflower seed, and especially with Niger seed which closely resembles thistle seed.

Chickadees

When you hear the winter woods ring with raspy little voices calling "chick-a-dee-dee-dee, chick-a-dee-dee-dee," you won't need eyesight to recognize chickadees. The "chick-a-dee-dee-dee" call will give you positive identification. But you shouldn't rely only on that call to identify chickadees, because chickadees have a whole variety of whistles and other calls that they might be using instead. You can easily identify chickadees by their coloring and behavior.

Chickadees are silvery, sparrow-size birds with dark caps on their heads and black bibs under their chins. Except for their caps and bibs, chickadees are a combination of gray and white. Two of our six species of chickadee also have some brown coloring on their sides or backs.

Have you ever seen a bird hang upside down on a twig? Chances are it was a chickadee or a relative of the chickadees. Chickadees are real acrobats, quickly and actively flitting around on tree trunks and small branches while searching for tiny insects. Chickadees can't stand upside down on tree trunks, as do nuthatches, but for these nimble acrobats hanging upside down from a twig appears to be just as easy as perching right side up.

You'll find winter the best time of year for seeing chickadees, because they move about in flocks then. The flocks are noisy with calls that help members stay together. The wide-ranging black-capped chickadee gives the easily recognized "chick-a-dee-dee" call the clearest, but all chickadees give some version of it.

Look for chickadees in shrubs and trees in any type of forest, as well as in residential areas that have lots of trees planted. Chickadees are frequent visitors to winter bird feeders, especially feeders filled with sunflower seeds or suet.

Chickadees are quite brave around humans and will frequently let you approach within several feet. Even with close inspection you may not be able to identify the particular species of chickadee. Even they can have a hard time telling each other apart.

Juncos

In winter juncos are one of the most common visitors to bird feeders. Since juncos are ground feeders, you'll often see flocks of them on the ground under the bird feeder. Most juncos summer and nest in the coniferous forests of mountains and the far north. When winter comes the juncos move down out of the conifer forests to look for food in fields and thickets, city parks, and suburban areas.

Juncos are dark, gray or gray-brown, sparrow-size birds with white breasts and white outer tail feathers. The key to their identification is the white stripes on the outside of an otherwise dark tail. The stripes are very easy to see when juncos fly. A few other small birds also have white streaks in their tails, but they don't have dark bodies with white stomachs.

In many parts of the country juncos are seen only in winter. This has given rise to the nickname "snowbird." You probably won't be able to find a listing for snowbirds in large bird books, but you may find it interesting to look up juncos. Five species of junco used to be listed—slate-colored, white-winged, Oregon, gray-headed, and Mexican—but recent books show only two species: dark-eyed and yellow-eyed.

Have whole species of juncos died off? No—but in 1973 ornithologists decided that since four of the former "species" of junco could interbreed, they must really be one species with four races, or, as scientists say, subspecies. The Mexican junco is still a separate species, but has been renamed yellow-eyed junco. The yellow-eyed junco lives only in Mexico and portions of southern Arizona and southern New Mexico, so all the juncos you see in the rest of North America are dark-eyed juncos.

The different races of dark-eyed junco are colored differently, though all have the identifying white breasts and stripes. In the East all of a junco, except the white breast and white outside tail feathers, is a uniform slate-gray color. Juncos that summer in the Black Hills are also slate gray, but they have two white stripes on their wings. In the West the slate-gray color often gives way to a variety of grays, dark browns, and even cinnamon brown. Western juncos often have gray heads and brown backs and sides.

Nuthatches

Nuthatches are one of the few birds in North America that can walk headfirst down a tree trunk. That's about all you'll need to remember to identify a nuthatch, for not only can they walk head down but they often do. They'll zigzag every which way over tree trunks and branches, but walking headfirst down a trunk is their preferred way to search for insects hiding in the bark. It's a trick that helps them find insects overlooked by all the heads-up woodpeckers.

Nuthatches are tree birds and can be found in all types of forests as well as orchards, residential areas with lots of trees, and even city parks if there are enough trees. The easiest place to find nuthatches is at winter bird feeders that offer sunflower seeds or suet. Nuthatches that come to feeders often become quite bold around people, and will let you get very close. Watch a nuthatch at a sunflower seed feeder and you'll see how they got their name, which originally was nuthack: They will fly away with a seed, wedge it in a crevice, and then hack it open with their bills.

Nuthatches are stocky little birds with short tails. They are smaller than sparrows, but stocky enough to sometimes give the impression of being larger. All four of our species of nuthatch have dark caps, white faces, and blue-gray backs, wings, and tails. The red-breasted nuthatch has a reddish breast. The white-breasted nuthatch has a white breast, as do the other two nuthatches. White-breasted nuthatches have black caps, which can help you differentiate them from the brown-headed nuthatch in the East, which obviously has a brown cap, and the pygmy nuthatch in the West, which has a gray-brown cap.

The white-breasted nuthatch is our most common and wide-ranging nuthatch, preferring deciduous or mixed woods. The red-breasted nuthatch prefers the northern and mountainous conifer forests. The brown-headed nuthatch likes loblolly and other pines of the Southeast. And the pygmy nuthatch prefers ponderosa and other pines of the West.

White-breasted nuthatches stay as mated pairs all year long. As they forage around through the woods and hedgerows, white-breasted pairs keep in touch by a nasal "yank yank" call that sounds very much like the squeak of two large tree branches rubbing together in the wind. Birds frequently mimic common background noises—it's a great way for two birds to maintain contact without attracting the attention of predators.

Downy and Hairy Woodpeckers

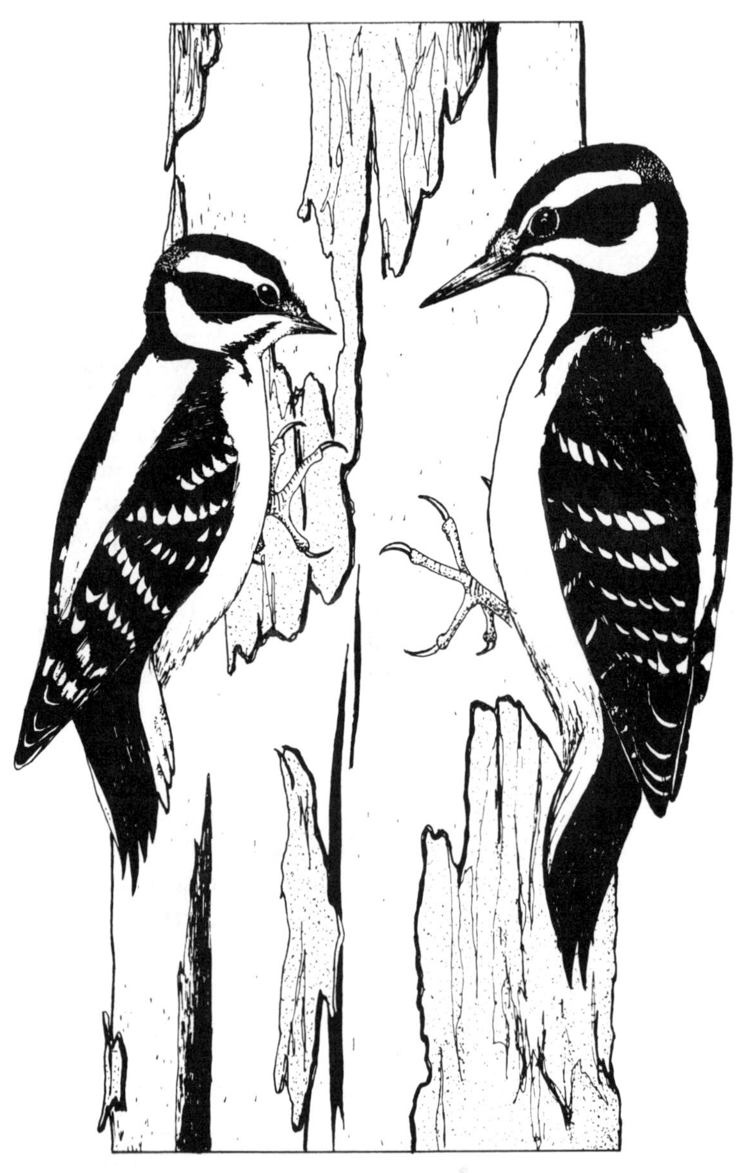

Downy and hairy woodpeckers are two of North America's most common woodpeckers, and they are colored almost exactly the same. You can identify a bird as either a downy or a hairy woodpecker by its principally white back. No other tree-trunk-clinging, woodpecker-type birds have white backs. Some have white rumps, but not white backs. Male downy and hairy woodpeckers have a little red patch on the back of their heads. Other than that, downy and hairy woodpeckers are black and white.

The easiest way to tell downy and hairy woodpeckers apart from each other is by their size, as the texture of their features, from which they got their names, is not that distinct except at very close quarters. Downys are only slightly larger than sparrows, while hairys are almost robin-size. Also, downy woodpeckers have small bills for woodpeckers. Their bills are about one third the length of their heads or shorter. Hairy woodpecker bills are quite large, about two thirds the length of their heads or longer.

You can find downy or hairy woodpeckers any time of year in practically any place with trees, from wilderness forests to city parks and suburban areas. You can frequently attract these and other woodpeckers to backyard bird feeders by hanging up some suet.

If you take a walk through woods in winter, you'll sometimes see bits of bark, often mixed with wood chips, on the snow under a tree, evidence of a woodpecker hard at work. When you watch a downy woodpecker in action you'll notice that it usually does not do much hard pecking. Instead, it pokes and peeks and pries, flipping off little bits of bark and lichen to expose insects and insect eggs. Hairy woodpeckers mainly go after wood-boring grubs by pecking holes through the bark and into wood borers' tunnels. Then they reach down the tunnel with their long, barbed tongues and pull out the insect. Both downy and hairy woodpeckers also eat some berries, seeds, and acorns.

Redheaded Woodpeckers

Redheaded woodpeckers are robin-size birds that can be easily identified by their completely red heads and necks. They are the only birds east of the Rocky Mountains that have solidly red heads and no red coloring anywhere else. From the neck down, redheaded woodpeckers are black and white. Their breasts are white, and their backs, wings, and tails sport a bold black-and-white pattern that is unique to them.

Like most woodpeckers, redheaded woodpeckers routinely fly in an undulating, roller-coaster style that gives you a good view of their back, tail, and wing patterns when they're flying away. Basically, redheaded woodpeckers appear black with a large white rump patch that extends out onto the wings. This pattern is so easy to recognize that you can easily identify young redheaded woodpeckers with it even though they are brown and white. Young redheaded woodpeckers don't start growing red head feathers until late in their first winter.

Redheaded woodpeckers live only east of the Rocky Mountains. You can find them in farm country and open woodlands, in city parks and golf courses, in orchards, in pastures with a few trees, and even out on nearly treeless prairies. Redheaded woodpeckers are willing to live where there are few trees because they don't do a great deal of wood pecking.

Redheaded woodpeckers do some foraging on trees, particularly dead trees, but they have many other sources of food. They rummage around on the ground after millipedes, centipedes, and other ground insects. They catch flying insects in the air, much as a kingbird does. They eat the eggs and young in other birds' nests. Redheaded woodpeckers also eat wild and domestic berries, cherries, and grapes. In the fall redheaded woodpeckers eat a lot of acorns, beechnuts, and corn, which they store for winter in any handy cavity. You can attract redheaded woodpeckers to backyard bird feeders with suet, sunflower seeds, cracked corn, or even old bread.

Northern Flickers

Almost half of the flickers' diet is made up of ants, so most of the flickers you'll see will be hopping around on the ground, where most of the ants are. Flickers flying up from a lawn after a meal of ants can be positively identified by the conspicuous white rump patch. Between a flicker's brown back and dark tail is a large white patch that shows clearly when it flies away in the undulating, roller-coaster style of flight so typical of woodpeckers. Flickers are a little larger than robins. The back of their wings is the same barred brown color as their backs. Flicker breasts are covered with round black spots, and they have distinct crescent-shaped black bibs. Flickers are woodpeckers and therefore have long woodpecker bills and stubby woodpecker tails.

The best time to find flickers is during the summer. Look for them in suburban areas and city parks, in farmland and lightly wooded regions. You'll find them in trees and more commonly on the ground. If you startle a flicker into flight and identify it by the flash of its white rump, you may find it interesting to walk over to where the flicker was on the ground and investigate what it was eating. Often you'll find a large anthill right where the flicker was standing. Flickers snatch up ants with their long, sticky tongues. They also eat just about any other ground insect or grub that they find, as well as some seeds and berries, including poison ivy berries.

Flickers can be noisy birds, particularly in the spring. They got their name from one of their calls, which sounds a little like "flicker, flicker, flicker." You won't be able to identify the call, however, just by the way it is written and pronounced. When you hear flickers give their "flicker" call, you'll realize that they pronounce "flicker" quite differently than we do.

Waxwings

Crested, black-masked, and grayish brown, waxwings are among the most beautiful birds in North America. They are larger than sparrows and have a very upright posture when perching in trees or shrubs. The easiest key to positive waxwing identification is the yellow band running across the tip of their tails. Waxwings are rather stocky and have a silky quality to their feathers that makes them seem very sleek. Their coloring enhances this sleek appearance. Waxwings aren't one or two solid colors. Rather, they are a wonderful blending of many different shades, from dull yellow-brown to gray.

Waxwings are tree birds. You'll find them in open woodlands, residential areas, and city parks. Waxwings are also very gregarious and social. Except for a while during nesting season, you'll almost always see waxwings in flocks of from four or five birds to over a hundred. You'll understand waxwing movement best if you think of them as gypsies. A flock of waxwings will suddenly show up one day, eat what fruit they can find, and just as suddenly move on. You can seldom predict their arrival, as you can with most birds.

There are two very similar-looking waxwings in North America, the cedar waxwing and the Bohemian waxwing. A difficult but reliable way to tell them apart is to look at the body feathers under their tails. These feathers are nearly white on a cedar waxwing and cinnamon-colored on a Bohemian waxwing. Bohemian waxwings are generally western birds and prefer coniferous forests. You can find cedar waxwings throughout southern Canada and the lower forty-eight states, in open deciduous woods as well as coniferous. The name cedar comes from their fondness for cedar berries, though they are also very fond of chokecherries and mountain ash berries.

Why the name waxwing? Well, certain feathers on a waxwing's wing are often tipped with shiny red spots that look like drops of old-fashioned sealing wax. Waxwings are fairly tame around people, and will let you approach close enough to get a good look at them, but you still may need binoculars to be able to see the tiny globs of "wax" on a waxwing's wing.

Eastern Kingbirds

Eastern kingbirds are the only North American songbirds with a broad white band across the tip of their tails. They are larger than sparrows and smaller than robins, with whitish breasts and black or dark gray heads, backs, wings, and most of their tails.

Despite what their name might imply, eastern kingbirds aren't limited to the eastern part of North America. They can be found as far west as British Columbia and central Oregon. As a matter of fact, about the only parts of the United States in which eastern kingbirds aren't found are California, Arizona, portions of New Mexico and Texas, and the extreme West Coast.

Look for kingbirds in open or partially open country around farms and orchards, in suburbs, at the edge of woods, and along the shores of lakes and rivers. You'll usually see them sitting out in the open on fences, electric lines, and dead branches sticking out from bushes or trees. Eastern kingbirds can be seen from late spring to early fall. The best time to look for them is in late August or early September, when they are flocking prior to migrating to South America.

Eastern kingbirds have a rather regal bearing and a little red crown patch on the top of their heads. But their name comes from their behavior, not their appearance. They act like petty tyrants in their nesting territory. Kingbirds chase all sorts of birds and animals away from their domain. Crows and hawks are special targets, but kingbirds will even chase eagles. The eastern kingbird's scientific name reflects this domineering nature: *Tyrannus tyrannus*.

If you watch eastern kingbirds during their nesting season, sooner or later you're bound to see them chase some creature. A typical chase begins with the kingbird sitting on an exposed perch scanning the air for flying insects. It will fly out on agile wings, grab an insect out of the air with its bill, and return to its post. When a crow comes into its territory, the kingbird takes off and flies above the crow. The kingbird then dives down, pecks at the crow's back, and flies up again. This is repeated again and again until the intruder is out of the kingbird's territory. And these aren't mock battles—kingbirds really peck. Eastern kingbirds have even been seen landing on a flying crow's back and pulling out a feather.

Wrens

With nine species of wren in North America, you'll be able to find wrens in swamps and deserts, in fields and forests, in wild, brambly thickets and manicured suburbs, and in city parks. Wrens are great and boisterous singers during their nesting season, so the best time of year to look for wrens is in early summer. All their chittering and singing can help you find them despite their fairly drab coloring. Wrens usually carry on their search for bugs fairly close to the ground, so look for them on the ground, in shrubs, and on low tree branches.

Wrens are very active and energetic little birds that are even smaller than sparrows. All nine species are brown or brown-gray and striped or spotted with black or white. They can usually be told from sparrows, warblers, and other small birds by their habit of holding their tails straight up. Wrens actually keep their tails down like other birds most of the time, raising their tails mainly when they are excited or disturbed. Fortunately for ease of identification, just the presence of a person close enough to identify the wren is often disturbance enough to raise a tail. The wren will frequently let you get pretty close, but its stubby little tail may be sticking straight up!

The cactus wren of the southwestern desert is an exception among North American wrens. It doesn't normally raise its tail, and it is larger than a sparrow rather than smaller. Cactus wrens do, however, share a lot of the other characteristics that make wrens so delightful to watch. They sing frequently and heartily, though not always melodiously. They are quick, active, and alert. Yet they allow people to approach quite close. They'll even nest close to houses, and you can often attract them by hanging a wren house 6 to 10 feet above the ground near some shrubbery that they can use for protection on their way to and from the nest.

HELPFUL TOOLS

Binoculars and Other Tools

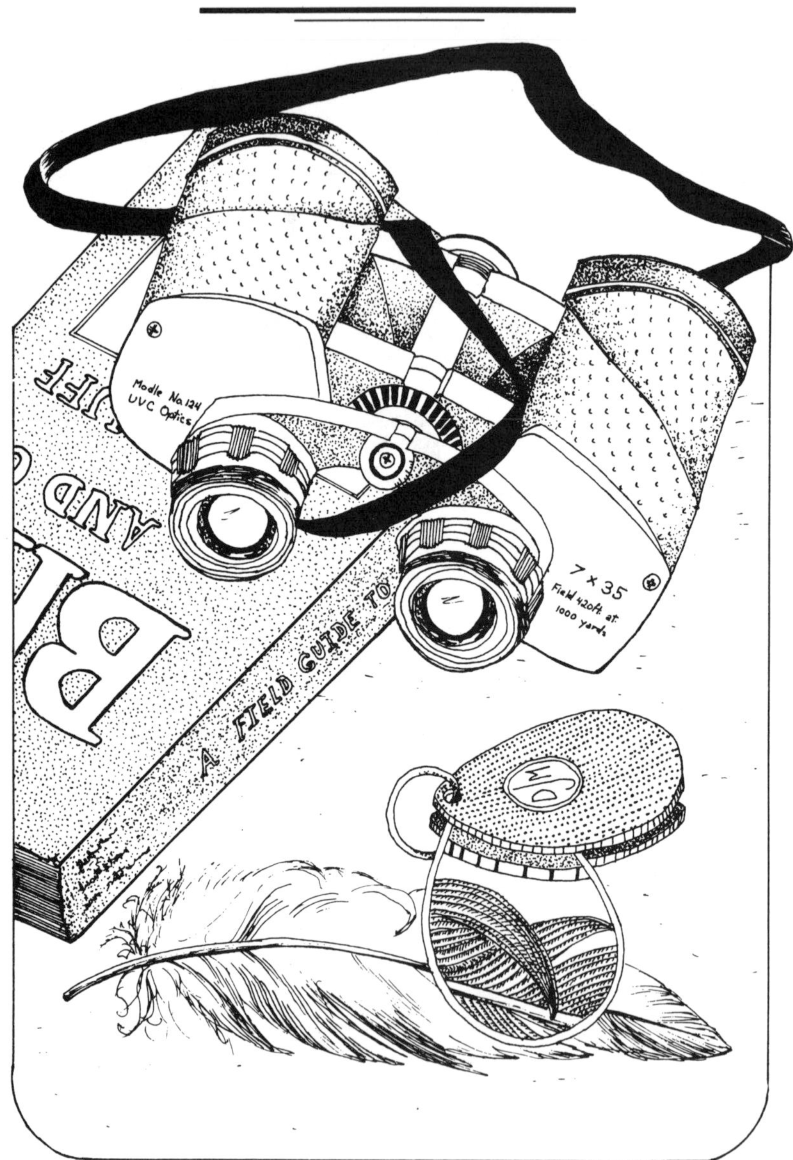

You don't need expensive tools to study nature. Combine your senses of touch, smell, and sight with a sense of wonder and you'll have it made. Some tools are great fun, though, and they can help you explore and learn more about the natural world around you than you ever could learn on your own.

Binoculars

Binoculars are helpful tools and the ones most commonly used by naturalists. They make distant objects appear larger and closer, allowing you to see them more clearly. The key to choosing the right pair of binoculars is to test how clearly you can see distant objects. You can do this right in a store by trying to read price tags or other small print on the other side of the showroom.

Binoculars are labeled with numbers, such as 7 × 35. The first number is the magnification power. Seven- and eight-power binoculars are probably best for all but the most experienced—binoculars with considerably higher power are harder to hold still long enough to get a good view, because every little jiggle you make is magnified greater. You'll see this right away when you try to read a price tag on the other side of a showroom with a high-powered pair of binoculars.

The second number is the front lens diameter in millimeters. Larger front lenses mean brighter images at the same power. With seven-power, 35mm lenses are common because they aren't too bright when you get some glare off snow or water, yet are still adequate at dusk. The field of view is also given for each set of binoculars, and also appears in terms of figures, for example, 420 feet at 1,000 yards. The wider the field of view, the easier it will be to locate birds or animals.

Telescopes

You'll find that telescopes and high-powered spotting scopes are of limited use for observing birds and animals, partly because they have such narrow fields of view that locating the bird or animal can be quite difficult. If you can keep the scope from wiggling too much, perhaps by bracing it on a tripod, it can be great for looking at things that don't move very much, such as bird nests or water birds swimming offshore.

Most scopes designed for land observation usually range between twenty- and sixty-power (magnification sixty times that of the naked eye), because distortions in the image caused by unevenly heated air near the Earth's surface show up too much under higher power. Telescopes with higher power are usually designed for stargazing.

Magnifying Glasses

Not all interesting sights are far away. Many of the most fascinating are close at hand but very tiny. This hidden world can be revealed with a relatively inexpensive magnifying glass. There are many different five-power, ten-power, or even higher-powered magnifying glasses on the market that are compact and sturdy enough to be taken out into the field. You can use them to peer closely into the structure and workings of insects, flowers, fallen bird feathers, and a myriad of other subjects.

Butterfly Nets

Butterfly nets aren't just for butterfly collectors. They can be used for everything from catching tadpoles for close examination to catching large June beetles at night to feed to orphaned baby screech owls.

Butterfly nets should be durable and large enough to fold over and contain a captured butterfly. Unfortunately, many are either expensive or difficult to find, or toys. One solution is to make your own out of mosquito netting, which you can buy at camping or sport stores. Any stick or dowel about two feet long will do for the handle. The hoop can be made out of heavy wire, such as a metal coat hanger, bent into a circle about one foot in diameter. To attach the wire hoop to the stick you may have to wrap the wire around the end of the stick, tie it snugly with string, and then tape it on as well. Sew the net about two feet long so that after you swoop at an insect you can turn the net sideways and have a large section in the end sealed off.

Cameras

"Take only pictures, leave only footprints" is an environmentalist's ideal. And you don't need fancy cameras to take good photographs of nature. Simple cameras can work well if you put extra effort into composing interesting shots.

More expensive 35mm cameras have two major advantages over simple automatic-focus cameras. First, you can control the light. This can be important in some pictures. For example, if you wanted to take a picture of a tree against the sky, an automatic

camera would average the light and give you a dark silhouette of the tree against a moderately light sky. If the light setting were adjustable, you could set the light properly for the tree and get a picture in which details of the bark were more visible, though the sky would wind up appearing much brighter.

Second, you can add a variety of lenses to many 35mm cameras: long telephoto lenses that make distant objects appear larger, short macro lenses that let you take close-up pictures of small objects, or macro zoom lenses that can do both. With a macro lens, you can have a single wildflower fill the whole frame.

Books

Most avid bird watchers find a detailed bird guide essential. Several very good bird guides, with color drawings and short descriptions of every bird that lives in or even visits North America, are on the market. Some birders also find recordings of bird songs helpful as well as interesting. Detailed field guides to mammals, butterflies, wildflowers, and other aspects of nature vary considerably and need to be chosen carefully to match up with your needs and interests.

Index

A

Air masses, and effect on weather, 95,
    101–2
Alto clouds, 97
Altocumulus clouds, 97
Altostratus clouds, 97
American coots, 139
American goldfinches, 153
American kestrels, 129
American lotus, 25
Ants, 87
Aphids, 77, 87
Aquatic animals
    beavers, 17
    nutrias, 17
Aquatic plants
    American lotus, 25
    coontail, 21
    duckweed, 23
    rushes and bulrushes, 27
    waterlilies, 25
    watermeal, 23
Armenian legend of the Milky Way,
    115
Atlas, and the myth of the Pleiades,
    113

B

Banded woolly bear caterpillars, 91
Barred owl, 121
Bats, 7, 125
Beavers, 17
Bees, 87
Beetles, whirligig, 73
Belted kingfisher, 136–37
Betelgeuse (star), 111
Big Dipper, 109
Big duckweed, 23
Binoculars, choosing, 175
Bird guides, 177
Birds, 119–71
    American kestrels, 129

belted kingfisher, 136–37
brown-headed cowbirds, 147
Canada geese, 143
chickadees, 47, 155
chimney swifts, 125
coots, 139
crows, 135
Eastern kingbirds, 169
goldfinches, 153
horned larks, 149
juncos, 157
mallard ducks, 27, 141
meadowlarks, 151
mourning doves, 127
nighthawks, 122–23
Northern flickers, 165
nuthatches, 155, 159
owls, 11, 121
red-tailed hawks, 133
swallows, 125
swifts, 7, 125
turkey vultures, 131
waxwings, 167
woodpeckers, 161, 163
wrens, 171
Black-and-yellow garden spiders, 81
Blackbirds, 143, 147
Black-capped chickadee, 155
Black widow spiders, 79
Bloodroot, 39
Bluebells, 51
Bobcat, 9
Bohemian waxwing, 167
Books, 177
Brown-headed cowbirds, 147
Brown-headed nuthatch, 159
Buckhorn, 59
Buffalo, 147
Bulrushes, 27
Buriats of Mongolia, and legend of the
    Milky Way, 115
Butterflies
    sulphur, 83
    swallowtail, 85
Butterfly nets, 177

C

Cactus, prickly pear, 55
Cactus wren, 171
Cameras, 177
Canada geese, 143
Canis Major, 111
Carrots, 53
Caterpillars
  banded woolly bear, 91
  fall webworms, 89
  of sulphur butterflies, 83
  of swallowtail butterflies, 85
  tent, 89
Cats, and cat tracks, 9
Cedar waxwing, 167
Celestial River, 115
Centipedes, 75
Chickadees, 47, 155
Chicory, 51
Chimney swifts, 125
Chipmunks, 5
Cirrocumulus clouds, 97
Cirrostratus clouds, 97, 105
Cirrus clouds, 97
Clouds, 97
Cold fronts, 102–3
Columbine, wild, 45
Common mullein, 57
Common nighthawk, 123
Common plantain, 59
Conifers, 37
Constellations
  Big Dipper, 109
  Canis Major, 111
  Lepus, 111
  Little Dipper, 109
  Milky Way, 114–15
  Orion, 111
Continental polar air mass, 95
Coontail, 21
Coots. *See* American coots
Copperheads, 3
Coral snake, 3
Corona, 105
Cottonmouths, 3
Coyote, 9
Crows, 135
Cuckoos, 89
Cumulonimbus clouds, 97
Cumulus clouds, 97

D

Daddy longlegs, 77
Damselflies, 69
Dark-eyed junco, 157
Deciduous trees, 37

Democritus, 115
Dogs, and dog tracks, 9
Dog Star, 111, 117
Downy woodpecker, 161
Dragonflies, 69
Ducks, 21, 23, 139
  mallards, 27, 141
Duckweed, 23

E

Eagles, 129
Eastern kingbirds, 169
Eastern meadowlark, 151
Eastern white pine tree, 37
Echolocation, 7, 73
English plantain, 59

F

Falcons, 129
Fall webworms, 89
Field guides, 177
Fish-eating birds
  coots, 139
  kingfishers, 136–37
Flannelleaf, 57
Flickers, 165
Floating heart, 25
Flowering plants
  American lotus, 25
  duckweed, 23
  prickly pear cactus, 55
  See also Wildflowers
Froghopper, 67
Fronts, warm and cold, 102–3

G

Geese, Canada, 143
Giant puffball mushrooms, 35
Goldenrod, 61
Goldenrod gall (insect), 64–65
Goldfinches, 47, 153
Grasses, 29
Gray-headed junco, 157
Great Dog constellation. *See* Canis
  Major
Great horned owl, 121, 135
Great Nebula of Orion, 111
Great Rift, 115
Groundhog, and spring predictions, 95
Ground squirrels, thirteen-lined, 5

H

Hairy woodpecker, 161
Harvestmen, 77

Hawks, 122, 129, 135
  red-tailed, 133
Hayfever, 61
Hibernation, 15, 91
High-pressure systems, 99
Hognose snake, 3
Honeydew, 87
Horned larks, 149
Horned owls, 11
Horsetail, 31, 33

I

Insects, 63–91
Iris, wild, 51
Isabella moth, 91

J

Jack-in-the-pulpit, 41
Juncos, 157

K

Kestrels. *See* American kestrels
Kingbirds. *See* Eastern kingbird
Kingfishers, 136–37

L

Ladybird beetle, 87
Ladybugs, 87
Larks. *See* Horned larks
Lepus, 111
Lesser nighthawk, 123
Lightning, 97
Little Dipper, 109
Long-eared owl, 121
Low-pressure systems, 99
Lunar halos, 105
Lynx, 9

M

Magnifying glasses, 177
Mallard ducks, 27, 141
Maritime tropical air mass, 95
Mayapples, 43
Mayflies, 71
Meadowlarks, 151
Mexican junco, 157
Milky Way, 114–15
Millipedes, 75
Molting of insects, 71
Mountain lion, 9
Mourning doves, 127
Mushrooms, puffball, 35
Muskrats, 17, 21, 25, 27

Mythology, and stars
  Big and Little Dippers, 109
  Milky Way, 115
  Orion, 111
  Pleiades, 113

N

Nighthawks, 122–23
Nimbostratus clouds, 97
Nimbus clouds, 97
Northern flickers, 165
North Star, 109
Nuthatches, 155, 159
Nutrias, 17
Nut sedge, 29

O

Opossums, 13
Oregon junco, 157
Orioles, 89
Orion, 111, 113, 115
Owls
  barred, 121
  horned, 11, 121, 135
  long-eared, 121
  western screech owl, 121

P

Pigeons, 127
Pine trees, 37
Piñon pine trees, 37
Pit vipers, 3
Plants, 19–61
Pleione, and the myth of the Pleiades,
  113
Pointer Stars, 109
Poisonous insects
  black widow spiders, 79
  centipedes, 75
Poisonous snakes, 3
Polaris, 109
Prevailing westerlies, 101
Prickly pear cactus, 55
Puffball mushrooms, 35
Pygmy nuthatch, 159

Q

Queen Anne's lace, 53

R

Rabbits, 11
Raccoons, 14–15
Ragweed, 61

Rain
   cause of, 99
   clouds and, 97
   fronts, warm and cold and, 103, 105
   lunar halo and, 105
Rattlesnakes, 3
Ravens, 135
Red-breasted nuthatch, 159
Redheaded woodpeckers, 163
Red-tailed hawks, 133
Rigel (star), 111
River of Light, 115
Roman legend of the Milky Way, 115
Rushes, 27

S

Scarlet king snake, 3
Scouring rush, 31
Sedges, 29
Silica, 31
Silver River, 115
Sirius, 111, 117
Slate-colored junco, 157
Snakes, 3
Snow
   cause of, 99
   clouds and, 97
   lunar halo and, 105
   spring snowstorms, 95
Snowbirds, 157
Sparrow hawks, 129
Sparrows, 147, 149
Spiders, 77
   black-and-yellow garden, 81
   black widows, 79
Spittlebugs, 66–67
Spring, air wars during, 95
Spruce trees, 37
Squirrels, thirteen-lined ground, 5
Starlings, 75
Stars
   Betelgeuse, 111
   Pleiades, the, 113
   Polaris, 109
   Rigel, 111
   Sirius, 111
   Venus, 117
Stinging nettle, 47
Stratocumulus clouds, 97
Stratus clouds, 97
Sulphur butterflies, 83
Summer, and stargazing, 115
Swallows, 125
Swallowtail butterflies, 85
Swift, 7

T

Taurus, the Bull, 113
Telescopes, 175
Tent caterpillars, 89
Thunder, 97
Toads, 75
Tracking animals
   cats, 9
   dogs, 9
   opossums, 13
   rabbits, 11
   raccoons, 14–15
Turkey vultures, 131

U

Ursa Major and Minor, 109

V

Velvetleaf, 49
Venus, 117
Vireos, 147
Vultures, 131

W

Warblers, 147
Warm fronts, 102–3, 105
Wasps, 67
Waterlily, 25
Watermeal, 23
Water shield, 25
Water striders, 73
Waxwings, 167
Weather signs, 93–105
   clouds, 97
   highs and lows, 99
   prevailing westerlies, 101
   spring air wars, 95
   warm and cold fronts, 102–3
Weeds
   classification of, 59
   common mullein, 57
   common plantain, 59
   Queen Anne's lace, 53
   stinging nettle, 47
   velvetleaf, 49
Western meadowlark, 151
Western screech owl, 121
Whirligig beetles, 73
White-breasted nuthatch, 159
White-winged junco, 157
Whorls, defined, 21
Wild columbine, 45

Wildflowers
  bloodroot, 39
  chicory, 51
  common mullein, 57
  goldenrod, 61
  jack-in-the-pulpit, 41
  mayapples, 43
  Queen Anne's lace, 53
  velvetleaf, 49
  wild columbine, 45
Wind, movement of, 101

Winter, and star gazing, 111, 113
Woodpeckers, 65, 153, 159
  downy, 161
  hairy, 161
  northern flickers, 165
  redheaded, 163
Wrens, 171

Y

Yellow-eyed junco, 157